Muscles, Molecules and Movement

Muscles, Molecules and Movement

An essay in the contraction of muscles

by J. R. Bendall
B.A. (Cantab), Sc.D. (Cantab)
Meat Research Institute, Langford, Bristol

HEINEMANN
LONDON

American Elsevier Publishing Company, Inc.
New York

ISBN (Heinemann) 0 435 62054 1
ISBN (American Elsevier) 0-444-19734-6
Library of Congress Catalog Card Number 76-88618

First Published 1969
Reprinted 1970, 1974

Published in Great Britain by
Heinemann Educational Books Ltd
48 Charles Street, London W1X 8AH

Published in the United States by
American Elsevier Publishing Company
52 Vanderbilt Avenue
New York, New York 10017

Printed in Great Britain by
Butler & Tanner Ltd, Frome and London

To

Φ.M

Drum geb' ich gern ihm den Gesellen zu,
Der reizt und wirkt und muss als Teufel schaffen.

GOETHE: *Faust*

Contents

II. INTACT MUSCLE

Acknowledgements

I owe a great debt to Dr Dorothy M. Needham, Prof. D. R. Wilkie and Dr H. E. Huxley for reading through the manuscript so carefully and making many helpful suggestions, and to Prof. R. E. Davies for many informative letters on the theoretical aspects of contraction. While thanking them warmly, I do not wish it to be thought that they necessarily concur in any of the controversial opinions I have expressed.

Particularly warm thanks are due to Miss Gwen Cowling and Mrs Daphne Hodgson for spending so much time and labour in typing the first draft of an untidy manuscript, and above all to my chief assistant, Mr C. C. Ketteridge, not only for the many drawings and diagrams, but also for carrying out so untiringly and exactly the experiments reported in chapters 2 and 3.

I am most grateful to Dr H. E. Huxley for allowing the use of the excellent photos in the early chapters; to Professor Jean Hauson for plate I.1c; to Dr Hasselbach for plates IV.2 and 3; Dr J. P. Revel for plate IV.1 and Dr G. Goldspink for plate V.1. I am also indebted to Professors A. V. Hill, Jean Hanson, A. F. Huxley and X. Aubert, and Drs J. Gergely and F. N. Briggs for helpful suggestions made in letters.

Last, but not least, I wish to acknowledge most heartily the invaluable and often acerbic advice of my old friend and colleague, Dr J. Thomas, particularly in matters English.

List of Plates

Introduction

'. . . jumping o'er times;
Turning th' accomplishment of many years
Into an hour-glass: for the which supply,
Admit me Chorus to this history;'

When the reader thinks of muscular contraction he will more than likely imagine to himself the actions of a ballerina, or a weight lifter, or a horse or greyhound running, rather than let us say, the beating of his heart or the slow movements of his gut. Even less will it occur to him to place the creeping of a snail or the closing of the shell of a mussel or the grabbing tentacles of a sea-anemone in the same category, yet all of these as well as the movements of many lowly animals and plants involve very similar, and sometimes even more highly organized, fundamental processes. Even among the higher plants, we can instance the dramatic movements of the flycatcher orchid, when it grasps its prey, or those of the leaves of mimosa when they are touched. And amongst less highly developed organisms, the undulations of the trypanosome of sleeping sickness, or the thrashing tails of myriads of spermatozoa during the fertilization of an egg, or the movements of chromosomes on the nuclear spindles of that egg, during its first mitosis and all the subsequent ones which have resulted in there being a reader for this page [160].

One feature common to all these diverse phenomena is the transduction of the chemical energy of ATP (adenosine triphosphate) into mechanical energy through the mediation of a system of contractile filaments; another common to many is the presence of two parallel sets of micro-filaments of the proteins, actin and myosin [160, 78, 57, 130, 140]. The specialized system of sliding,

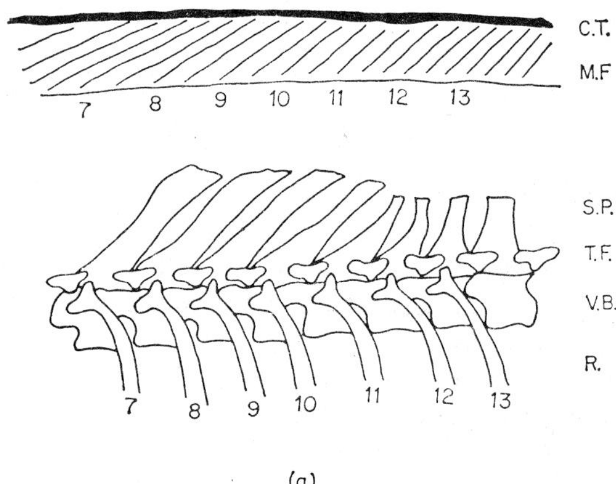

Figure A.i. (a) Diagram of part of longissimus dorsi muscle of ox, to show direction of fibres in relation to the spine. CT = connective tissue sheath, MF = muscle-fibres, SP = spinous processes, TC = transverse processes, VB = vertebral bodies, R = ribs, with numbers. Connective tissue or tendon shown in all figures as thick black lines.

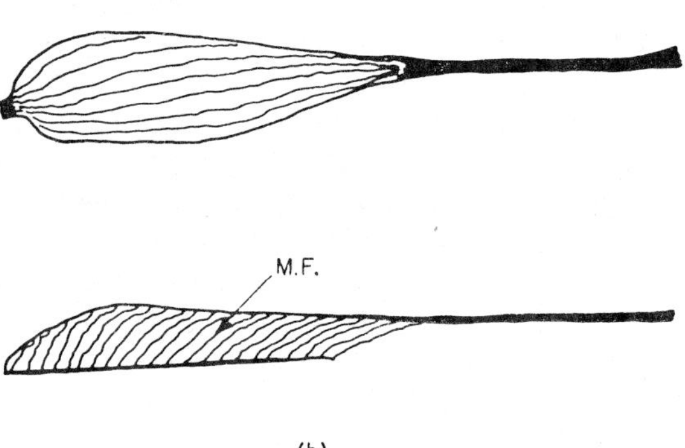

(*b*) Flexor carpi radialis muscle of ox, before and after cutting through it longitudinally. Note the short fibres, in the cut section, running from the ventral to dorsal aspect of the muscle. The muscle itself is 19 cm long.

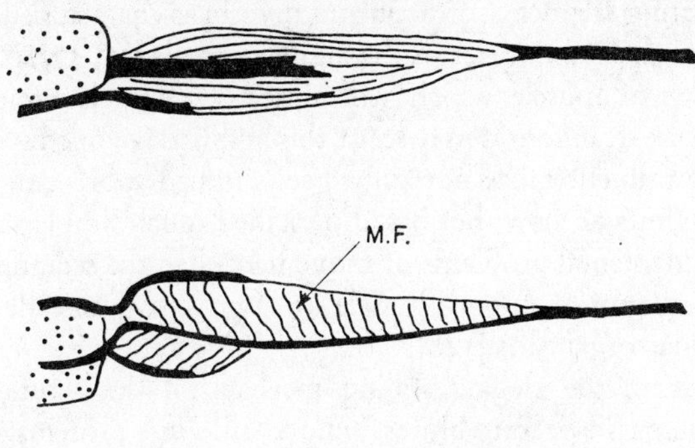

(c)

(*c*) Semitendinosus of rabbit, 3·5 cm long, before and after cutting in the same manner as in (*b*).

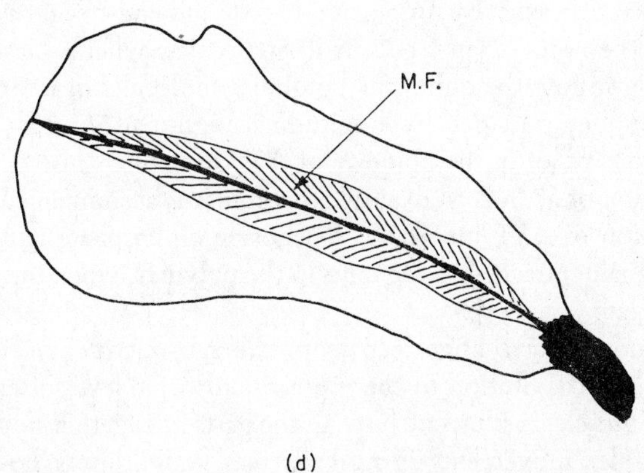

(d)

(*d*) Typical bipennate muscle, infraspinatus of the ox, cut open along its length, to show muscle fibres running from the outer connective tissue sheath to the thick internal sheath, which runs down the middle of the muscle. Length of muscle 39 cm.

(All drawings in figures A.i to iv by Mr C. C. Ketteridge, Meat Research Institute, Langford, Bristol.)

interdigitating filaments of actin and myosin is characteristic of all the striated skeletal muscles of the higher animals [56, 57, 80, 81]: it is this type of muscle which will almost exclusively concern us here, because strange as it may seem, this highly developed system of energy transduction has not only been studied more extensively from all points of view, but has also yielded much simpler answers to the fundamental problems of movement than the seemingly less complex systems of smooth muscle or of primitive unicellular and multicellular organisms [140].

Apart from the elegant sliding mechanism, the actin–myosin system presents a feature almost unique amongst proteins: it is at once structural, in the sense that it runs continuously within the fibres of a muscle from one end to the other, and enzymic, in the sense that the energy for the sliding movements comes from the splitting of ATP at enzyme sites on the heads of the myosin molecules [46] and the subsequent transduction of this energy into movement through interactions with the closely neighbouring actin filaments [144, 145]. Hence myosin acts as an enzyme, in monomer or polymer form, and is also capable of forming very long, well-organized filaments, by simple side-to-side and end-to-end aggregation of the non-enzymic tails of its molecules; whereas actin is an example of how the quite small globular molecules of a monomer can form long polymers by longitudinal alignment. And if we wish for further proof of the ubiquity of ATP in nature, this very polymerization of actin is an example, because it is accompanied by the conversion of ATP into ADP (adenosine diphosphate); the latter itself remains attached to the units in the polymer, when the process is complete [57, 143].

We now take a huge leap from the sub-microscopic level of molecular organization to the supra-microscopic level of an intact skeletal muscle, to illustrate how all the parts fit together to produce the complex movements of our ballerinas, weightlifters, horses and greyhounds, which in themselves are neither more nor less beautiful than the molecular interplay on which they depend.

Figure A summarizes the main structural features of muscle [58]. The first part shows four types of muscle (figure A.1). The first (*a*), the longissimus dorsi of the back, is the longest muscle of the body, but this does not mean that the fibres and bundles of fibres of which it is composed have to run the whole of this length. On the contrary, they arise from spinous processes, and in an animal such as the ox, run forward and downwards from there to the transverse processes, and to the lateral surfaces of the ribs. The muscle is covered dorsally and laterally with a thick sheath of collagen, from which many of the fibres originate. Owing to this structure, none of the fibres need be much more than 12 cm long, even in this very large animal. The same applies to the second type, called fusiform muscles, in which sheets of connective tissue often continue over the surface of the muscle from the tendons of origin and insertion, and the fibres can be attached to these at a sharp angle; this reduces their required length: examples are the flexor carpi radialis of the ox (*b*) and the small semitendinosus of the rabbit (*c*). In the third type, called bipennate (*d*), this economy in length of fibres is taken even further, for the fibres run in at an angle from the collagenous sheets on the surface to other parallel sheets within the muscle. This type of structure is taken to extremes in multipennate muscles such as the large deltoid muscle of the upper back, where internal collagenous sheets divide the muscle into three or more parts. In a fourth type exemplified by the long semitendinosus muscle in the hind limb of the ox, the fibres appear to run the whole length of the muscle, in this case about 47 cm, but they are strengthened by many parallel elastin fibres.

The contractile fibres of muscle usually end on connective tissue, which is continuous with the tendons at each end, and bears the tension of contraction. This connective tissue is mostly collagen, the fibres of which are almost steel-like in their strength and exhibit very little 'give' when under tension: for example the enormously strong Achilles tendon in the heel, which in its upper part gives off branching sheets of connective tissue which partly cover the soleus and gastrocnemius muscles of the calf. The larger connective tissue sheets, or fasciae, of the muscles give off finer and ever finer branches which penetrate between the muscle substance, in the form of a kind of connective tissue stocking, the threads of which form a complex three dimensional weave, without interfering with the longitudinal alignment of the muscle fibres themselves. It is impossible to illustrate this structure diagrammatically, but the cross-section of a small part of a muscle (figure A.ii) shows how the collagen network becomes finer the closer it gets to the muscle fibres themselves; the latter are usually arranged in primary bundles of 20 to 40 fibres.

The structure of a fibre is shown in longitudinal section in figure A.iii, and in cross-section in figure A.iv. Figure A.iii shows the

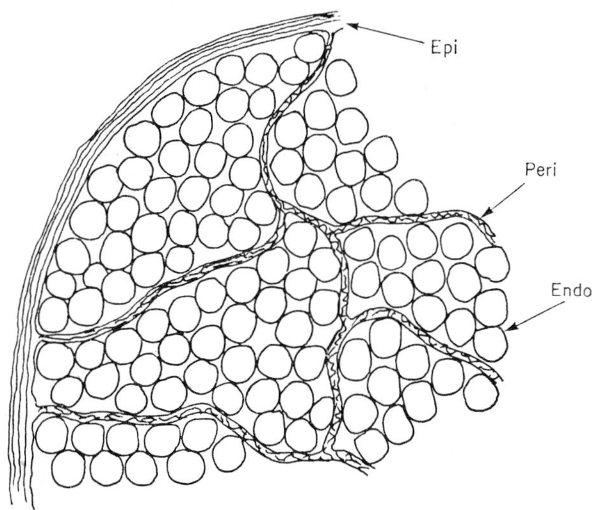

Figure A.ii. Cross-section of muscle fibre bundles near the surface of a muscle. On the outside can be seen epimysium (epi) which binds the bundles together; individual bundles are surrounded by perimysium (peri), and individual fibres by endomysium (endo). The connective tissue consists mainly of collagen, with a few elastin fibres. Diameter of muscle fibres = about 50 μ.

characteristic cross-striated pattern which we shall discuss in a moment. The very fine line on the outside of the fibre, in both figures, is supposed to represent the true cell-membrane or plasmalemma, which like the membranes of all other cells is in fact triple-layered, and about 100 Å thick. The fibre itself is between 20 and 80 μ in diameter (0·02 to 0·08 mm or 200,000 to 800,000 Å). Within the fibre immediately under the plasmalemma, nuclei occur about every 5 μ along its length (these are not shown). There are many nuclei because the long thin muscle fibre is developed from

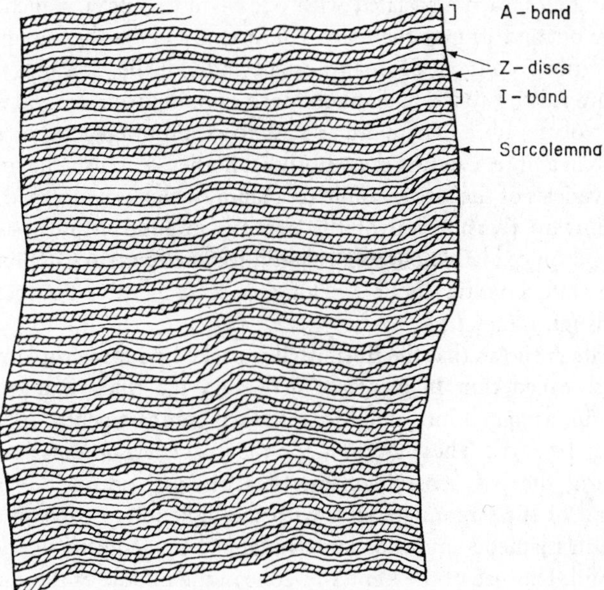

Figure A.iii. Longitudinal section of a skeletal muscle fibre to show cross striations. Note the bounding sarcolemma, part of which is the thin plasmalemma proper. Under the sarcolemma many nuclei, about 2 μ long, are found (not shown). Diameter of fibre about 50 μ.

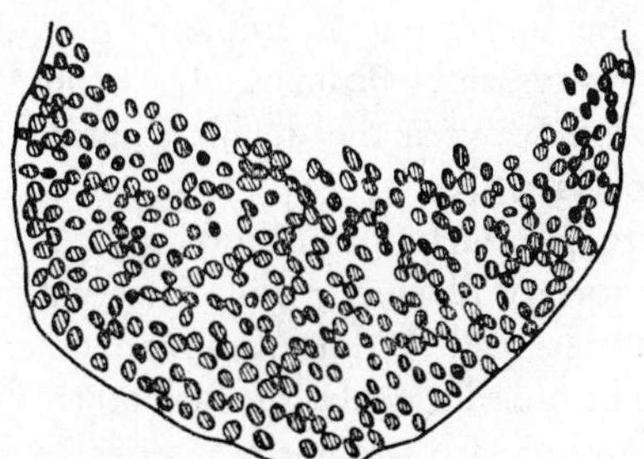

Figure A.iv. Cross-section of part of a muscle fibre, to show fibrils packed within. The packing in unfixed fibres is much tighter than that shown here. Running between the fibrils would be the transverse tubules of the sarcoplasmic reticulum, which cannot be seen in the type of preparation shown here. Diameter of fibrils = 1 to 1·5 μ.

myotubules, each associated with one or more nuclei, which later move out and arrange themselves in rows just under the membrane. Each fibre consists of many small individually striated fibrils (figure A.iv), 1 to 2 μ in diameter, each running from one end to the other of the fibre. There are at least 1000 and often 2000 or more fibrils in a fibre. Coating these fibrils is another structure, the tubules and vesicles of the sarcoplasmic reticulum, which is concerned with the inward transmission of the nervous impulse. Muscles which depend on oxidative phosphorylation for their energy supply, also have mitochondria about 2 μ long, packed in rows between the fibrils (cf. figure 4.1).

Plate A shows that the fibril itself is packed with microfilaments, the so-called thin filaments of actin and the thick filaments of myosin, arranged into small longitudinal compartments, or sarcomeres [57, 58]. These account for the characteristic cross-striated pattern; the less dense filaments of actin give rise to the 'light' I-band of the sarcomeres seen with the light microscope, and the myosin filaments, more or less overlapped by actin, to the denser A band. The size of the lighter H-zone in the middle of the A-band depends on muscle length, that is, on how far the actin filaments happen to have been pulled out from between the myosin filaments, with which they interdigitate. The actin filaments themselves arise from the Z-discs, clearly seen as a thin line in most photographs taken with a light microscope. These discs contain other proteins besides actin: the major one is tropomyosin, which resembles myosin in many ways, and which possibly extends from the Z-discs in either direction between the double stranded helices of actin monomers (molecules) [37, 51, 81].

It does not require much imagination on the readers' part to see how these two sorts of filaments, while preserving their individual initial lengths, can be made to slide over one another as the muscle is stretched, or to be pulled actively past each other, the actin moving towards the centre of each sarcomere in series, as the energy from the splitting of ATP is transduced into movement.

Part One: Muscle Proteins, Fibrils and Fibres

I: Structure and Organization of the Contractile Proteins

Myosin, actin and tropomyosin

The proteins directly involved in the contraction of striated muscle, myosin and actin, are organized in the muscle fibrils in the form of two sorts of filaments, thick and thin (plates I.4*a* and *b* and

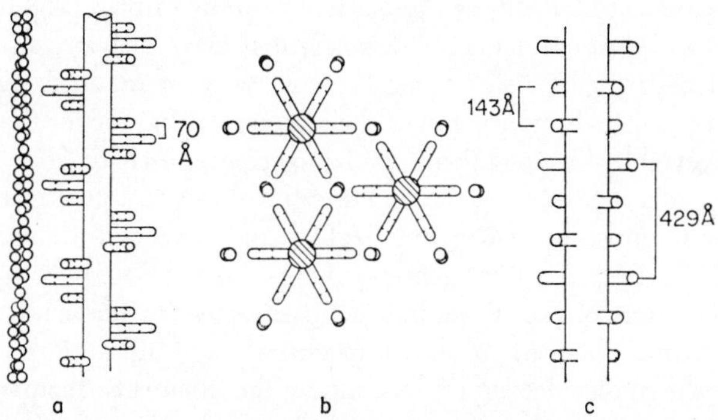

Figure I.1a and b. Diagram to show structure of actin (thin) and myosin (thick) filaments. Note (a) the double-stranded super-helix of actin, the pitch of the spiral lying between 350 and 400 Å (see text); and the six staggered rows of heads on the myosin filament; (*b*) the alignment of one actin filament opposite each row of feet, in cross-section. (After [82].)

Figure I.1c. The structure of the thick filaments of muscle, as shown by recent X-ray analysis [83]. Note that, unlike the model in (a) deduced from EM evidence, the X-ray model has pairs of heads, each on either side of the filament.

figures I.1*a* and *b*) [56, 57, 58, 81]. It is the thick filaments, each consisting of either 180 or 360 longitudinally oriented molecules of myosin, which give rise to the high density and characteristic anisotropy of the A-bands of the sarcomeres of the muscle, whereas the much less dense isotropic I-band is made up of the bead-like molecules of the smaller globular protein, actin, arranged in double-stranded chains [81, 121]. The total number of beads of actin in each thin filament in a half sarcomere is 400, or 200 in each strand of the chain. This gives a total of about 800 molecules of actin from tip to tip of a single chain, passing through the *Z*-disc (figure 1.3). In the sarcomeres of the higher vertebrates the chemical evidence favours a molar ratio of actin to myosin of approximately 4 to 1, and a weight ratio of 1 to 2, and this would require the presence of at least 360 molecules of myosin in each thick filament of a sarcomere [58, 63, 83, 122].

The reality of the two sorts of filament, and the conclusion that muscle contraction and relaxation consist of the active or passive sliding of one sort of filament over the other, rests not only on the electron-microscope pictures of intact sarcomeres, but also on the fact that intact filaments can be isolated from myofibrils [81]. From plate I.4*a* or figure 1.1*a*, for instance, it is clear that any reagent which destroys bonds between the two sorts of filament should cause complete disintegration of the sarcomeres. In practice this can be achieved by homogenizing a small piece of muscle to fibril level in a blender, and placing the resulting fibrils in a so-called relaxing medium consisting of Mg ions, ATP† and the Ca-ion chelating agent, EGTA†. Further gentle homogenization causes complete disintegration of the fibrils into actin and myosin filaments. The EM pictures obtained are shown in plates I.1*a*, *b* and *c*.

Of the molecules which constitute the filaments themselves, (plates I.1*a* and *b*), that of myosin is long and thin, with a thickened portion at one end. It has a total average length of about 1500 Å and a breadth of about 20 Å in the rod-shaped region [81, 133]. The thickened heads of the molecules are 200 Å long by about 40 Å in breadth. From these values, the axial ratio is nearly 100, which is considerably higher than the value of about 25 deduced from hydrodynamic studies in solution.

† ATP = adenosine triphosphate; EGTA = ethylene-glycol-bis(-amino-ethyl-ether)-N, N′-tetraacetic acid.

The actin molecules, about 55 Å in diameter as monomers, are arranged like the strings of beads in a necklace, but double-stranded, with the strands twisted round one another in the form of a super-helix (plate I.1*c* and figure 1.1*a*). We use this latter term to avoid confusion with the α-helical or double or triple-helical configuration of the polypeptide chains of the proteins themselves. These double super-helices of actin monomers have a fairly regular pitch which it is often difficult to appreciate from single photographs. However, judged by large numbers of such photographs, the cross-over points of the two strands evidently lie between 340 and 420 Å, depending on the method of preparation, but apparently regardless of species [57, 59, 112]. It is difficult, however, to eliminate the shrinkage of the EM preparations, due to fixation, so that support has been sought from low and moderate angle X-ray diffraction studies of unfixed preparations or of living muscle itself [44, 59, 83, 112]. Even here it is still difficult to make a final decision, though the most recent evidence [83] strongly suggests a value near to 360 Å. Although the filaments have a regular pitch, this does not necessarily mean that there should be an integral number of beads in that pitch. However, taking 55 Å as the diameter of an actin monomer, the pitch of the helix would be 710 Å for thirteen beads, giving a value of 355 Å for the distance between cross-over points, which is very close to the value given by X-ray analysis and falls within the range of the EM measurements [83, 59].

The measurements are complicated by the fact that striations often appear in natural and artificial thin filaments at about 400 Å in the electron microscope [35, 45], possibly corresponding to a meridional reflection in the X-ray diagram at about 385 Å. It is not yet known for certain what this spacing is due to, but the most recent evidence suggests that it comes from tropomyosin, possibly associated with a newly discovered protein, troponin, both of which are orientated within the grooves of the actin helices. However, we may safely conclude that the actin monomers of the thin filaments are arranged in double super-helices with a pitch between 710 and 740 Å [44, 83].

Sub-structure

The primary structure of myosin and tropomyosin is characterized by a high content of acidic and basic amino-acids, which confer a very high charge on the molecule (see table 1.1). Actin, in contrast, has a rather low charge, and is further distinguished by its high proline content. This imino-acid, when present in a polypeptide chain, makes a kink in it, so that the formation of long stretches of α or other helical confirmations is difficult. This, combined with the effect of the large number of bulky non-polar side-chains, is sufficient to explain the very low helical content of this protein, found by optical rotatory dispersion (ORD) measurements. Indeed, if the proline were equally distributed along the chain, we should expect stretches of about twenty amino-acid residues between each proline kink, and these would be in the form of more or less extended polypeptide chains, bending backwards and forwards over each other, with the non-polar groups on the inside of the molecule and the charges on the outside. The form of the molecule would be spherical, with the structure stabilized, in the absence of any measurable S—S links between 'cysteine' residues, by the non-polar groups on the inside and by interchain hydrogen bonding between peptide bonds, and also between the uncharged polar groups; this would give a very compact structure, with no water trapped within it. The diameter of such a molecule might be expected to be about 60 Å, allowing 3 Å as the distance between neighbouring peptide bonds. This is not far from the actual value found by electron-microscopy of 55 Å [59].

Myosin as we have seen is quite distinct from actin and possesses long straight stretches of high helical content, which seem to extend for almost 1300 Å in the tail region [102, 103, 133]. The tropomyosin molecule is of similar construction but shorter [97, 102]. No doubt the formation of such structures is made possible by the interactions of the large numbers of negatively and positively charged side-chains along the molecules. In the tail region of myosin, for instance, there is a negative or positive side-chain for every third uncharged residue. And interestingly enough, there is almost no disturbance of the linearity by proline residues.

A feature of myosin not possessed by tropomyosin is, of course, the fattened head region. In plate I.1*a*, this appears as a relatively small portion of the whole structure, yet it is here that the enzymic and actin-combining sites are found, and where the transformation of chemical into mechanical energy takes place, when the molecules are in their orderly array in the thick filaments of the myofibrils. The nature of this head region has been partly uncovered by the massive chemical attack mounted against it during the last decade and a half. It was early discovered that trypsin breaks the molecule more or less specifically into two parts, the so-called light (LMM) and heavy (HMM) meromyosins; the former comes from the straight tail region, while HMM comes from the head and 'neck' regions, and possesses all the enzymic properties of the parent molecule [102, 146].

The light fraction (LMM) closely resembles tropomyosin B in structure, and many of the latest X-ray and optical rotatory dispersion measurements upon it suggest that it consists of two identical polypeptide chains coiled round one another in slightly distorted α-helical form, giving the so-called coiled-coil structure [102, 103]. It may, however, consist of a triple helix, on the evidence of the break-up of the molecule in 12 M urea or 5 M guanidine HCl (166). This gives rise to some difficulty in explaining the structure of the head region, which then would also have to consist of three polypeptide chains. We shall come back to a discussion of this aspect in a moment. There is general agreement, at least, that LMM is highly helical and straight and provides the backbone structure of the organized thick filaments of the A-band. In the EM its length appears to vary from about 600 to 900 Å which agrees well with hydrodynamic studies [81, 133].

The heavy fraction (HMM) is more complex than LMM, and contains not only the head and neck regions of the molecule, but also part of the tail [102, 146]. This is obvious not only from the physical measurements, but also from the EM photos. The latter give variable lengths, depending on where trypsin has attacked the parent molecule, which it can evidently do over long stretches of the backbone chains. The range is at least from 500 to 700 Å, which is a third to half the length of the entire molecule, whereas the fattened head region itself measures only about 200 Å (plate I.1*a*). From all these measurements, the salient fact emerges that HMM has a very much lower helical content than LMM, and that most of this helix is situated in the neck and the part of the tail still attached to it, about 400 Å in length [102]. Thus it is probable that the true head region contains little or no helix, but is constructed rather more like a typical 'globular' protein such as actin. This also follows from the very high proline content of HMM (see table 1.1) [114].

Further enzymic attack by trypsin on HMM yields so-called sub-fragment 1 (HMM S–1) which possesses all the enzymic and actin-combining properties of the parent molecule [114]. Under the EM, it appears as a circular object, smaller than the head of myosin proper [133]. Its molecular weight lies between 120,000 and 180,000 [122], the most recent value being 129,000 [114]. How much of the true head region such a molecule could account for depends on what molecular weight we choose for the entire head, but it now seems likely that it makes up about half. This conclusion emerges from recent EM studies in which myosin molecules were 'shadowed' by the method known as rotary shadow casting [141*a*]. In this method, the sample is rotated while being shadowed with gold or platinum, thus allowing much greater definition in the EM photos.

The results show clearly, first that the myosin molecule is double-headed, and that the size of the individual heads agrees very well with the dimensions of molecules of HMM sub-fraction 1; secondly, that the heads may be found lying either one above the other in the direction of the long axis of the molecule or side by side, showing that the short portion by which they are attached to the LMM neck and tail is extremely flexible; thirdly, from companion studies of the products of papain digestion, it is found that the portion of the tail (HMM sub-fragment 2) still attached to the heads in preparations of HMM, is highly helical and straight. The results can best be illustrated by a diagram (figure 1.2); its main features are that it can explain all the known particle weights of the various sub-fragments which also fit in satisfactorily from the point of view of length, and that it shows the tail region to consist of a double helix or coiled coil attached to the heads by extension of the individual elements of the coils. This short region between the heads and the tail consists of single polypeptide chains which because they are so obviously flexible are unlikely to be completely α-helical.

We can summarize the complex secondary and tertiary structure of the contractile proteins by saying that myosin and tropomyosin B are long molecules with axial ratios in excess of 20, capable of forming aggregates of 1 μ (= 10,000 Å) or more in length, by end-to-end and side-to-side electrostatic interaction between their tails; actin, on the other hand, is a globular molecule of much lower molecular weight, and only about 55 Å in diameter, which aggregates under physiological conditions into long double-stranded chains of beads

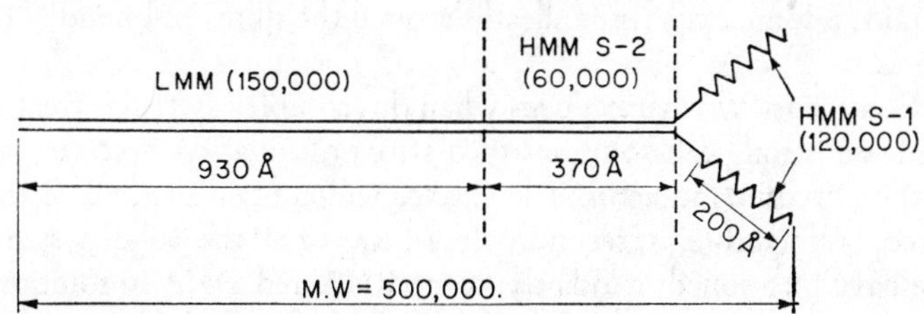

Figure 1.2. **Diagram of a myosin molecule, as visualized by EM photos taken after rotary shadowing of the specimen [141a].**
LMM = light meromyosin
HMM S-2 = heavy meromyosin sub-fragment 2
HMM S-1 = heavy meromyosin sub-fragment 1
(Note that the long straight portion of the molecule is shown here as two parallel straight rods, which in fact are twisted about one another to give a helical coiled-coil structure.)

of indefinite length, if the protein is pure. In the muscle itself, however, a limit of between 1 and 1·3 μ is imposed upon the length of a single actin chain, suggesting that some automatic coding is present in the developing muscle. Possibly the presence of tropomyosin and perhaps troponin provides the necessary code, but at present there is no clear evidence that this is so [51, 59, 102, 183].

Other structural proteins of the myofibril

When talking of the possibility of other structural proteins existing in the myofibril, we must become conscious of how all these entities are experimentally prepared and what is present to contaminate them. The most important contaminants arise from three sources: (i) the so-called soluble proteins of the sarco-plasmic fluid bathing the fibrils and containing a multitude of enzymes mostly involved in the glycolytic cycle; (ii) the mitochondria which lie between the fibrils and which are involved in the utilization of oxygen and the cycle of oxidative phosphorylation; and (iii) the tubules of the sarcoplasmic reticulum which delineate the fibrils themselves and which act as the transmission system for the nervous impulse. In addition, there is the possibility of minor, but important contamination, from the RNA of the ribosomes and the DNA from the many nuclei which lie just under the fibre membrane, and also from the connective tissue sheaths around the fibres and bundles of fibres.

Consider now what happens when this complex system is treated with the strong salt solutions of 0·5 to 1·0 molarity, necessary to extract myosin and actin. The extract will contain first, all of the sarcoplasmic proteins; secondly, fragments of all the 'solid' systems we have mentioned; and lastly, some RNA and DNA in solution. The next stage is to centrifuge out the larger non-soluble particles, and then to precipitate myosin or actomyosin, by diluting with water to the appropriate ionic strengths. In the course of this precipitation the nucleic acids are certainly carried down, and contamination from them is found in most myosin solutions. Moreover, unless great care is taken with the re-precipitation procedures, contamination from small particles of sarcoplasmic reticulum and mitochondria is unavoidable. Since actin itself is never extracted quantitatively by any of the procedures, a plethora of trace proteins have been reported as contaminants not only of actin and myosin preparations, but also in the residue left after their partial extraction. Therefore we must approach the problem of additional contractile components with the greatest caution, and not hesitate to demand the most stringent proof of a new entity. As examples of such entities, isolated in the past and since proved to have no real existence, we may quote X- and Y-proteins and 'contractin'.

In spite of the above criticisms, there is nevertheless evidence, not only from chemical investigations but also from low-angle X-ray studies [20, 35, 83, 122], that one or more structural proteins, besides actin, myosin and tropomyosin, are present in the sarcomeres, albeit in small quantities. Of these, α-actinin and troponin seem to be fairly well-defined, both of them resembling actin itself in many respects [35]. Troponin in particular has, like actin, a very high proline content, so that one would expect this protein to be globular rather than fibrillar, by the same arguments as we used in the case of actin itself. It has been suggested that α-actinin is present together with tropomyosin in the Z-disc, whereas troponin is associated with the actin filaments themselves, where like tropomyosin, it is thought to be present within the grooves of the actin superhelices.

Impure tropomyosin when added back to a purified system of actin and myosin confers upon the ATP-ase activity of the latter the extreme sensitivity to traces of Ca ions, which is such a marked feature of intact fibre and fibrillar preparations. Pure tropomyosin, even with its SH-groups carefully protected against oxidation, fails to do this, and it has recently been confirmed independently that the impurity which confers Ca-sensitivity on the system is troponin [20]. Since we have concluded that troponin is a globular protein, its role must evidently be to bind the long molecules of tropomyosin together within the actin helices. In fact, staining the sarcomeres with ferritin-labelled antibodies reveals that this troponin occurs in 'striations' every 400 Å or so along the sarcomere [35].

It has also been reported that troponin binds 5 Ca^{++} per 10^5 g, the binding constant being 6×10^5 M^{-1} [35*a*] in the presence of 4 mM Mg^{++}. Its molecular weight, from the sedimentation constant of 3 S, is of the order of 80,000 [35*a*], so that each molecule would bind 4 Ca^{++}; from this and its known spacing along the I-filaments, it follows that there is 1 molecule of it per 7 or 8 actins, that is, one molecule for every two or so myosins. In terms of myosin molecules, therefore, the apparent binding would be 2 Ca^{++} per molecule. As we shall see, this number is a highly significant one, when considered in relation to the known activating effect of Ca, in trace amounts, on native actomyosin systems which contain tropomyosin and troponin. It should also be noted that pure tropomyosin is itself incapable of binding Ca to any considerable extent.

In the arguments which follow we shall take the view that troponin occurs along with tropomyosin in the actin filaments, but because there is still some lack of definition, we shall refer to this complex as 'tropomyosin', thus leaving aside detailed arguments about its structure.

Before leaving this aspect of structure, we should mention another protein, myo-fibrillin, which has recently been extracted from residues [54]. This is supposed to be the so-called S-protein, originally proposed to join the actin filaments together in the middle of each half sarcomere. Such a protein would have to be extremely elastic, however, to account for the properties of muscle [58, 170].

Formation of actin filaments (thin)

The phenomenon of the aggregation of the globular molecules, or monomers, of actin, called G-actin when in this form, into the long two-stranded super-helices called F-actin (see plate I.1*c* and figure 1.1*a*), was one of the first properties of this protein to be recognized [143, 146]. This G–F transformation comes about in the following way during the course of preparation: myosin is first removed from a muscle mince by short extraction at high ionic strength; lipid is then removed from the residues by treating them with acetone, and they are then dried; the dried residues are extracted in water containing some ATP, and this breaks up the actin filaments, still partially organized in the myosin-free sarcomeres, into their separate beads which appear in the extract as G-actin. For this process to occur ATP is essential. Now, by raising the ionic strength ($\Gamma/2$) of the extract to 0·1 with KCl, the solution suddenly becomes more viscous, and if examined in the electron microscope is seen to conta in once more the characteristic double-stranded helices of F-actin, but now of very variable length. By keeping all the solutions at 0°C, it is possible to prepare tropomyosin-free F-actin by this method, but if the temperature is accidentally allowed to rise, more or less contamination with tropomyosin will result [33].

The chemistry of the G–F transformation is unique and takes place as follows:

$$n(\text{G-actin-ATP}) \rightleftarrows (\text{F-actin-ADP})_n + nP_i \qquad 1.1$$

The reaction is a purely stoichiometric one, in which the molecules of G-actin are transformed to the F- or chain-form, while at the same

Plate A. Longitudinal EM section of four frog muscle fibrils to show general arrangement of actin and myosin filaments. Distance between Z lines (dark cross-lines on plate) is about 2·5 μ. Note vesicles of sarcoplasmic reticulum at the z-band level. (Photo by courtesy of Dr H. E. Huxley.)

Plate 1.1a EM photo of individual my- sin molecules, by the shadow castir technique. The average length is abo 1500 Å. Note the globular head and long tails.

Plate 1.1b Preparation of A-filament from rabbit psoas muscle (negativ staining). The length of the intact fila- ments is about 1·6 μ, and there ar also some broken ones. The surface c the filaments is covered with smal projections, the myosin heads.

time the bound ATP is itself changed to ADP, also bound. There is a great loss of configurational entropy during the process, because the system passes from a state of random dispersal of G-actin molecules to that of their polymerization into an orderly array in the new super-helices of F-actin. The exact nature of the link between the actin monomers is not known, but it apparently does not involve the bound ADP which is such a characteristic feature of the native filaments, because this ADP can be removed by sonification without destroying the structure [5]. Aggregation may occur here in the same way as it does in the case of tobacco mosaic virus protein, that is by the removal of one molecule of water per mole from the so-called iceberg regions of the protein, and the interaction of this 'dehydrated' non-polar centre with that on neighbouring molecules [101]. It is now also certain that G-actin, free of bound nucleotide, can polymerize to the F-form, just as well as it does in the presence of ATP [5].

The equilibrium of the G–F transformation lies far to the right in equation 1.1, and so cannot be reversed to any considerable extent by adding an excess of P_i to F-actin-ADP, that is to say, it cannot act as a kind of ATP-resynthesizing machine [143, 146]. Reversal can in fact be achieved only by adding ATP to an F-actin solution at low ionic strength, and it is strictly speaking an exchange reaction, accompanied by a large increase in configurational entropy:

$$(\text{F-actin-ADP})_n + n\text{ATP} \underset{(\leftarrow)}{\rightleftarrows} n(\text{G-actin-ATP}) + n\text{ADP} \qquad 1.2$$

The nature of the reaction has been proven by the use of radioactively labelled ATP, and the same technique has been used to show that it does not occur to any considerable extent during the contraction of myofibrillar systems; during the supercontraction of artificial actomyosin, however, there is considerable exchange of labelled ATP with the bound ADP of actin, this newly bound ATP becoming rapidly dephosphorylated to ADP [146*a*]. Thus, although some exchange might occur between F-actin and ATP during the natural process of contraction, it is not likely that it takes place via reaction 1.2, particularly as the ionic strength of about 0·15 which obtains there is also too high to permit it.

These properties of actin suggest that its biosynthesis is relatively simple. It need only fall off the ribosome [120] as a polypeptide chain which automatically coils itself up into the G-form, because of the conditions, but then, finding itself in the presence of ATP, immediately polymerizes according to equation 1.1 into an orderly array of new super-helices of F-actin-ADP. We must, however, assume that tropomyosin, possibly together with troponin and α-actinin is present at the right time and place to act as a core to the F-actin filaments, and thus as a code to control their length [7, 11].

Formation of myosin filaments (thick)

The formation of the large filaments of myosin characteristic of the A-band of the sarcomeres can now also be achieved *in vitro* [81, 93, 94]. It is done by lowering, more or less quickly, the ionic strength of a solution of myosin, in 0·6 M KCl, to about 0·2. The first step is the uniting in pairs of the tails of a number of molecules to give a short filament with a bare central shaft, and with the heads of the parents protruding at each end of it, as seen in plate I.2. In its simplest form, this process goes on by addition of new molecules lengthwise in each direction, until a maximum length of about 1·5 μ is achieved, as seen in the plate [81]. By this time, the structure is very similar indeed to the thick filaments which can be directly isolated from myofibrils (see plate I.1*b*); these also have a bare central shaft about 1500 to 2000 Å long (= 0·15 to 0·2 μ), on either side of which regular projections of the heads of the myosin molecules occur right up to the ends of the filaments, which terminate in a gradual taper. The formation of this very orderly structure is highly dependent on ionic strength and pH [93, 94]. Outside the natural range of $\Gamma/2$ = 0·15 to 0·17, and pH = 7·2 to 7·5, aggregation often occurs in a very irregular way, giving rise to twig-like structures and very long aggregates.

The simplest explanation of the aggregation phenomenon is that the tails of the individual myosin molecules are attracted to one another by the high electrostatic charge they possess, and that the ionic bonds formed are stable at physiological ionic strength and pH. It is not, however, so easy to see why under certain conditions they form up in the polarized way they do, with a bare central shaft and the molecules on either side of it orientated in different senses [81]. What causes the first aggregation to be tail to tail, and all subsequent ones to be tail to head? Perhaps when we know more about the structure of myosin, and particularly the amino-acid sequence in the tail region, an explanation will present itself.

These remarks take us about as far as it is possible to go with the organization of molecules without looking in detail at actual photographs of the A-bands and I-bands of muscle. They satisfactorily establish that myosin molecules can aggregate in a polarized fashion to give thick filaments up to 1·5 μ long, and that these show the heads of the molecules protruding at intervals on either side of a bare central shaft in a very similar manner indeed to the thick filaments, directly isolated from muscle.

Myofibrillar organization

In the skeletal muscles of the higher mammals the unit of myofibrillar organization, the sarcomere, appears in the EM as shown in plates I.4*a*, *b* and *c* [56, 80]. Plate I.4*a* shows the thick myosin filaments of the A-band, with the heads of the molecules protruding regularly along their length in the form of cross-bridges to the actin, except in the central shaft-region in the M-band where the union of molecules is tail to tail. In the centre of the latter band the filaments appear thicker, and some other protein of unknown constitution may be present there to strengthen the union of the tails (plate I.4*b*). The thinner actin filaments, in this case in a partly stretched muscle, run from the Z-discs through the I-band and then between the myosin filaments, until they terminate just short of the M-band region in each half sarcomere. As we shall see, the actual point of termination of the actin filaments in the A-band depends solely on the length at which the muscle is fixed for sectioning. Most muscles can be stretched so far in the pre-rigor state that the actin filaments are pulled almost out of the A-band, whereas in a supercontracted muscle they can penetrate between their opposite numbers in the other half of the sarcomere, until the Z-discs come into contact with the ends of the myosin filaments and even begin to compress them [56, 57, 58].

We show a cross-section of a fibril in plate I.4*c* (cf. figure 1.1*b*), taken through a region of overlap of actin and myosin filaments in the A-band. This demonstrates the typical hexagonal arrangement of the two sorts of filament in this highly organized band; part of the section has been taken through the I-band where we see only rather less regularly disposed filaments of actin, whereas in the M-band region in another part of the section, we see regularity of the myosin filaments, but without actin filaments. The hexagonal array at the point of overlap of actin and myosin consists of six actin filaments surrounding each myosin filament; looked at the other way about, this means that each actin is surrounded by three myosins (cf. figure 1.1*b*).

The dimensions of the two sorts of filament in most striated muscles are: for actin, 2·2 μ from tip to tip through the Z-line, by about 80 Å in diameter; for myosin, 1·5 μ by about 120 Å [81, 119]. The heads of the myosin molecules protrude about 100 Å from the surface of the filament to form the so-called cross-bridges with neighbouring actin filaments, but the gap they must span is variable and determined by muscle length, as we shall see when we come to discuss centre-to-centre filament distances.

We must now enquire how the protruding heads of the myosin molecules are arranged around and along the thick filaments and how these are related to the surrounding actin filaments, something it was impossible to deduce from the isolated filaments of myosin we described in the last section. Careful measurement of many EM photos shows that a myosin head occurs about every 400 Å along any one row in a filament, and since from the cross-sectional appearance each thick filament is surrounded by six actin filaments, it is natural to conclude that there are six rows of such heads in the filament circumference, giving a six-fold screw axis of symmetry. Each head in a row would then be about 70 Å from its nearest neighbour in the next row. Such a structure is illustrated in figure 1.1*a*. Recent low-angle X-ray studies, using long tubes to give very high definition [83] have confirmed that there are indeed six rows of heads in the filament circumference; however, the occurrence of a very strong third-order meridional reflection at 143 Å makes it apparent that the axis of symmetry is not six-fold, but rather that the heads are arranged in pairs (one head on either side of the filament) at 143 Å intervals, with each pair rotated relative to its neighbours by 120° (figure 1.1*c*). The exact spacing between heads in any one row is then 429 Å, and each head sticks out 135 Å from the centre of the filament. Because of the way the filament is built up, with a tail-to-tail alignment of molecules in the central M-band region, and a head-to-tail alignment extending on each side of the latter, the polarity of the myosin heads is reversed at the M-line, corresponding as we shall see to the opposite polarities of the actin filaments in each half-sarcomere [81, 121]. Note that the thick filament contains 30 projecting heads in each of its six rows, giving a total of 180.

At first sight it appears simple to build up a filament from long molecules of myosin in the head-to-tail manner we have outlined above, but the reader can easily prove to himself how difficult this is by attempting to do so with matchsticks, at the same time keeping strictly to the spacings shown in figure 1.1*c*. He will very soon find himself in difficulties even with the very first row of molecules, if the heads are made to protrude correctly from the surface. By using fluorescent antibody staining it has indeed been shown that the required geometry is most complex [121]. Twelve molecules of myosin are required in the cross-section of the filament at the M-band level, that is, in the central tail-to-tail region, in order to give the correct distribution of heads in the more distal portions, and these are tightly packed in triangular shape. This results in the rest of the filament, apart from the tapered ends, also being tightly packed with myosin molecules with no free space within, and it further implies that every now and then a myosin head must be able to protrude from quite deep within the structure. Hence the neck region of the molecule must be fairly flexible, whereas the tail, or L-meromyosin region, must be rigid to withstand the longitudinal stresses in the sarcomere when it contracts. This indirect evidence that the neck region is different from the tail proper, and rather more randomly coiled is confirmed by the recent EM photos we discussed earlier, which show that individual myosin molecules are double-headed, each of the heads being attached by a single flexible polypeptide chain to the double stranded helices of the tail [141*a*] (see figure 1.2). According to this new model, the rather puzzling thickenings in the M-band region of tail-to-tail abutment of molecules are produced by the cross links between the myosin molecules themselves, possibly involving another protein.

Stoichimetry of actin and myosin in muscle

The exact stoichimetry of actin and myosin in the sarcomere is still the subject of debate. Chemical studies show that myosin makes up about 54 per cent by weight of the structural proteins of muscle, and actin about 27 per cent, the remainder mostly consisting of tropomyosin (about 12 per cent), and troponin and the two actinins (about 7 per cent). These figures are necessarily somewhat inexact, because none of the present methods gives complete extraction of any of the components so that roundabout arguments have to be used to assess their likely content [58, 63, 83, 122]. Nevertheless the molecular ratio of actin to myosin on the above basis could not be more than 4 to 1, whereas if we calculate from the known lengths of the thick and thin filaments in muscle and the number of molecules they apparently contain, we arrive at a much higher value of about 8·7 to 1 (counting 2 actin filaments per myosin, the former containing about 1600 molecules and the latter about 180 projecting heads). The latter ratio would require actin to make up 54 per cent of the structural proteins, which is impossible if the myosin content is to be taken as 50 per cent as well. The only simple way out of this difficulty is to assume that each projecting myosin head is in fact double, that is, it contains two molecules of myosin instead of one [cf. 83 and 121]. This is not a very attractive solution to the problem, because it results in a clumsy structure, particularly as each myosin molecule is itself double-headed [141*a*], but on present evidence it is the only way of resolving the difficulty.

The Z-disc and the polarity of actin

The Z-disc is the point at which the actin filaments from neighbouring sarcomeres are joined together. Its gross structure is best shown from EM photos of actin filaments, separated from muscle by the procedures we have outlined, but still attached to the Z-disc (plate I.3*a*) [81]. Here we see the filaments from neighbouring sarcomeres densely knotted together in the disc, but spread out freely and randomly on either side of it. To carry the argument further, use is made of an important property of actin and myosin systems, the ability of the two sorts of molecules to aggregate to form complex 'molecules' of actomyosin.

When actin and myosin combine to form actomyosin, they do so by the myosin heads attaching themselves, one to every two actin beads of the super-helix, to leave the long tails of the myosin molecules free, but helically arranged along the long axis of the actomyosin thread (cf. figure 2.4). Such a preparation of actomyosin, because of the tails which are about 1200 Å long, entangling themselves into complex ladder-like structures, does not give very clear EM photos, but if the myosin molecule is simplified by chopping off part of its tail by means of trypsin, and using the shorter H-meromyosin instead, a much more orderly array of the molecules results, without too much risk of side-to-side aggregation [81]. We then obtain the picture shown in plate I.3*b*. We can now see a very distinctive arrowhead formation of the H-meromyosin molecules along the actin helices, with the arrows pointing in one direction on one side of the dense Z-disc, and in the opposite direction on the other side. Hence, it is deduced that the actin helices have one polarity on one side of the disc and the opposite polarity on the other, and this must obviously apply also to the actin filaments in each half of an intact sarcomere. Together with the already demonstrated polarity of myosin in each sarcomere, this has important implications for the sliding filament model of contraction.

The detailed construction of the Z-disc itself is still a matter of dispute, but it is now justifiable to assume that the actin filaments from neighbouring sarcomeres are here linked together through the agency of a tropomyosin-like molecule, which starting from this point, runs in either direction in the groove of the actin superhelices. Strong support for this idea is given by the occurrence in tropomyosin crystals of a nearly square lattice in which two filaments cross at each lattice point. The dimensions of the lattice and its angles are almost identical with those found in cross-sections of the Z-disc [48, 81].

The sliding filament model of contraction

The sliding filament model of contraction is probably the most significant single contribution to the subject in the last decade, and if we base all our future discussion upon it, many of the tricky problems of energetics, for instance, will be found to fall into place [56, 57, 58]. In this chapter, we do not discuss the biochemical implications, nor indeed the underlying molecular mechanism, because they will be dealt with more exhaustively in their proper context, but rather give a general description of the process.

It is probably simplest to approach the model from the observations of phase-contrast microscopy (see figure 1.3). We see first that a muscle actively shortens by the Z-discs of the sarcomeres being drawn in towards the A-band, resulting in a narrowing of the I-band, but *without* any change of length in the A-band. Conversely the I-band appears to lengthen again when the muscle is restored to its original length after a contraction, but still without any change of length of the A-band. Secondly, by extending a resting muscle, a zone of lesser density appears on either side of the M-line, its width depending on the muscle length. This is called the H-zone. Thirdly, unloaded contractions of a muscle can reduce its rest length to 0·4 or less of the original value, and these excessive contractions are accompanied first by a decrease in width of the H-zone and then by an increase of density in the M-band region, which gradually spreads throughout the A-band, and finally, just after the complete disappearance of the I-band, by the appearance of a thickening in the Z-line region, commonly called a contraction node. The fourth and last point is that a muscle can pass into rigor and become rigid and inextensible at almost any length from very long to very short. Any forcible attempt to stretch it is then accompanied by fracture, usually in the I-band.

Hence, the material of the A-band must be a more or less rigid and inextensible structure, whereas the I-band appears from EM

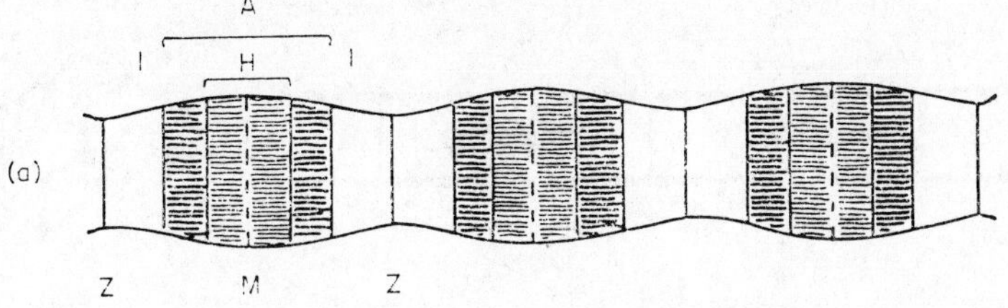

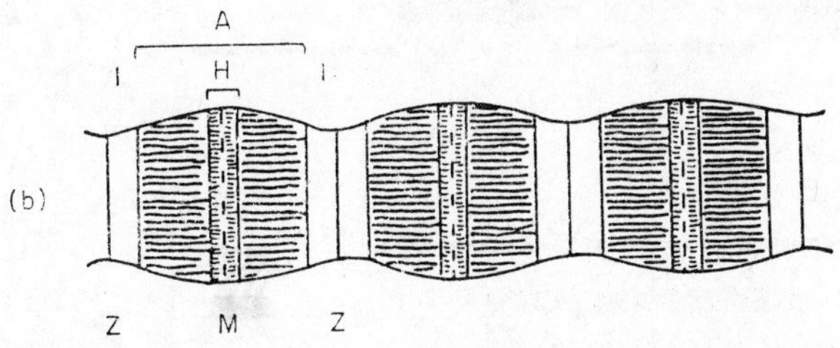

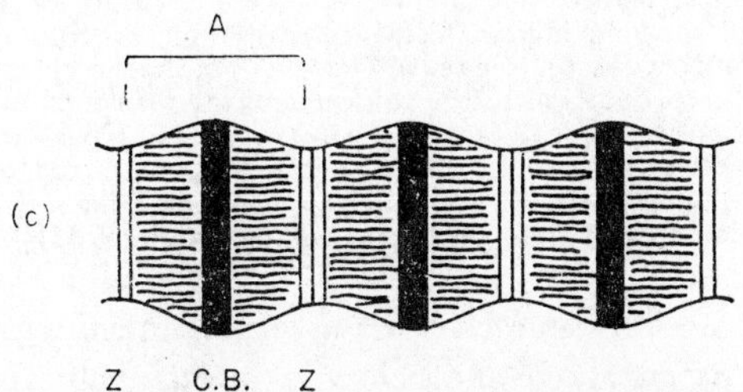

Figure I.3. Diagram to show the appearance of individual myofibrils in the light microscope, at three different sarcomere lengths. The diameter of the middle fibril of the three, at rest length (l_0), is about 1 μ. Note the contraction bands formed in the M-line region of the bottom fibril, when the sarcomere has shortened to 1·5 μ. Here $l_0 = 2{\cdot}2\ \mu$.

photos to be more flexible, but equally inextensible when pulled taut. Moreover, during contraction, considerable movement of material must occur to account for the density changes.

Now we return to the electron microscope evidence, as shown in plates I.4*a* and *b*. It is easiest to reproduce this plate in diagram form,

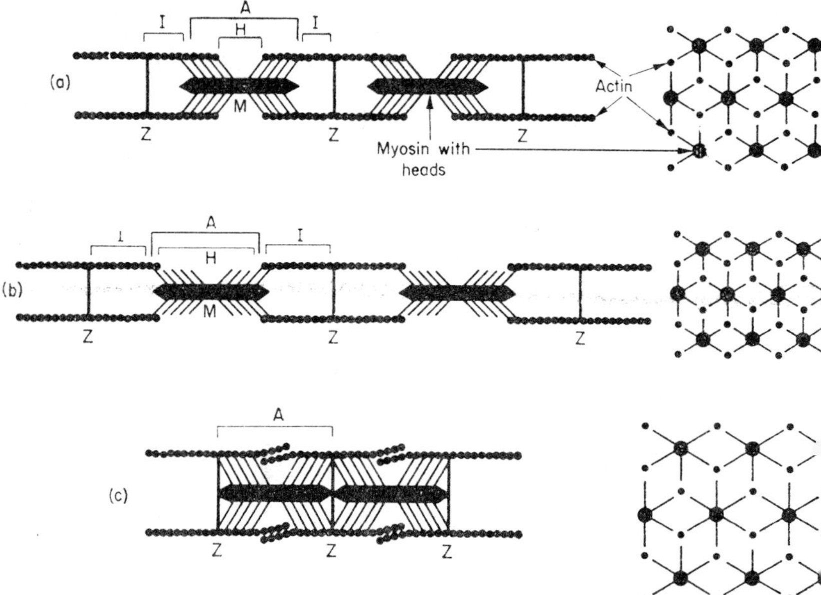

Figure 1.4. Diagram of fine structure of a sarcomere of ox muscle in longitudinal and cross-section, showing myosin filaments with their protruding heads, and the beaded structure of actin (see text). The length of the sarcomere has been reduced tenfold in proportion to the thickness of the filaments, for ease of drawing. *A*: sarcomeres at near rest length ($= l_0 = 2{\cdot}4\ \mu$) per sarcomere; *B*: sarcomeres after stretch to $1{\cdot}3\ l_0$; *C*: sarcomeres contracted to $0{\cdot}6\ l_0$. Note that the myosin heads are drawn sloping towards the Z-lines, the position in which they could exert their maximum pulling action (cf. [31]).

as in figure 1.4, where we show in each case a myosin filament with two of its six actin partners above and below it, in each half sarcomere. We picture the actin filament, like any other long protein filament, as a fairly rigid, inextensible chain, linked in series with the filaments in the next sarcomere through the Z-disc, also inextensible. Similarly, we picture the myosin filament as a rigid and inextensible rod, about twice as thick as the actin filaments, with the projecting heads of the molecules arranged along it. The lengths, in frog muscle, for instance, are $1{\cdot}0\ \mu$ for the actin filaments in each half sarcomere, and $1{\cdot}5\ \mu$ for the myosin. They are somewhat longer in the ox.

If we were to put the two sorts of filaments together in this orderly way, with say, only dilute salt present, then they would immediately form cross-bonds with one another, wherever myosin heads were in easy range of actin monomers. This could occur at any of the lengths shown in the diagram. The whole system would then be locked up from end to end in series, and would be inextensible. This we can consider as the *rigor state* [14, 130].

Now what happens when we add ATP to the system, in the presence of Mg ions? This depends very much on the exact conditions: in the complete absence of Ca ions, removable for instance by adding a calcium chelating agent, or alternatively in the presence of an SH-blocking reagent, the myosin ATP-ase sites on the heads can be completely inhibited and then MgATP acts as a plasticizer and prevents the formation of cross-links. In this state the actin filaments can be pulled past their myosin partners without any resistance: in other words, the whole system would then come completely to bits under stress, were it not for the presence, in the intact muscle, of connective tissue and other structures in parallel, to prevent this happening. We can picture a resting living muscle as being in this easily extensible condition, the only resistance to stretch coming from the parallel elements we have mentioned. This we will call *the resting state*. But note that when we speak of extensibility in this context, we do not mean that the rigidity of the individual filaments has in any way altered, but only that they can slide freely past one another, retaining their individual lengths.

If we now add to this hypothetical system a small trace of Ca ions, the immediate effect is to activate the ATP-ase sites on the heads of the myosin molecules; ATP is split and interactions between the heads and neighbouring actin monomers begin. The free energy released in this process is used for the development of tension, or for shortening and the performance of work. During the shortening, the individual filaments do not alter their length, but merely slide past one another (figure 1.4). How has this active sliding been effected? We still do not know precisely, but low-angle X-ray observations of living muscle show that there certainly is movement of the myosin heads, which could either consist of a waving to and fro, thus pushing the actin partner towards the centre of the sarcomere, or of a pulling action, due to miniature contractions of the head and neck regions of the individual molecules [83]. This state of the system, we shall call the *contractile state*. As we shall see, it can be reversed *in vivo* and *in vitro*, merely by removing all traces of Ca ions, providing ATP or some other 'plasticizing' polyphosphate compound is present.

The above hypothesis can now almost certainly be promoted to the status of a theory. It has been found to hold good in all the types of muscle where a double array of filaments is easily discernible, and provides an elegant and convincing explanation of all the facts of contraction, and also for the first time, of the development of rigor. As we have said, the latter state sets in when ATP is absent, and can occur at any length within the range of overlap of actin and myosin filaments.

The only problem is that of the so-called resting tension of muscle, that is, its capacity to recover from an imposed length change. Very probably this is brought about by the return to their original length of spring-like elastic elements in parallel with, and surrounding the individual fibrils and fibres which make up the intact muscle [124]. Alternatively, we could consider this as the function of the hypothetical S-protein which at one time was thought to connect the ends of the opposing actin filaments in each sarcomere. However, the presence of such a protein, unless it were extremely extensible and very thin, would be more of a hindrance than a help to recovery from stretch. The only evidence for it is indirect, in the sense that the 'extensible' protein, myo-fibrillin, recently isolated from muscle residues, might fill the bill [54, 170].

X-ray evidence for the sliding filament model

So far we have only discussed the microscopic evidence for the sliding filament theory. There is, however, a wealth of recent evidence from low-angle X-ray studies, the most important feature of which is that they can be carried out on living contracting muscle, or on unfixed preparations [43, 44, 82, 83].

The first point brought out by these studies follows from the long established fact that the volume of a muscle stays virtually constant, when it contracts. Hence, the diameter ($2r$) of any 'micro' cylindrical portion of the muscle we like to consider must vary inversely as the square root of its length (L), since $r = \sqrt{V/\pi L}$, where V, the volume, is constant. Further, since this micro-cylinder is filled with actin and myosin filaments, the actin-to-myosin filament, centre-to-

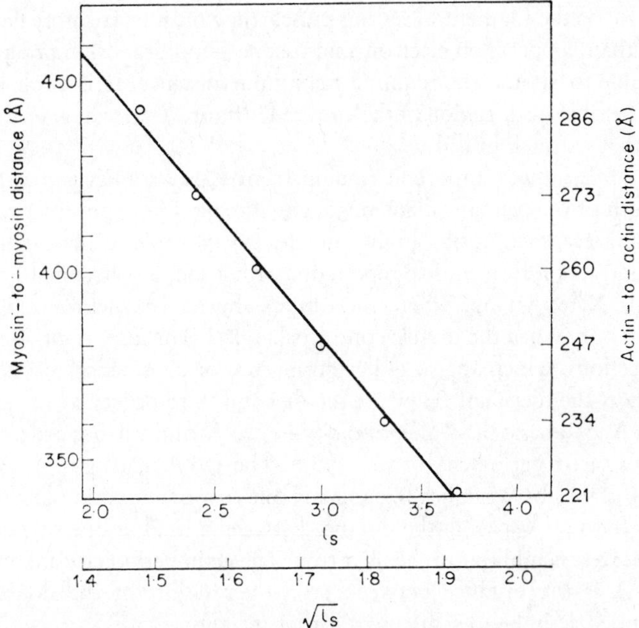

Figure 1.5. **Plot of the myosin-to-myosin and actin-to-actin centre-to-centre distance, as measured by X-ray diffraction, against the square root of the sarcomere length of a fibril ($\sqrt{l_s}$) (see [112]).**

centre distance, should also vary inversely as the square root of the length. This is entirely borne out by comparing actual measurements of equatorial X-ray reflections with the sarcomere length, in living muscle. The X-ray results are shown in figure 1.5, where we see that the square root of sarcomere length (l_s) plots linearly against the myosin-to-myosin centre-to-centre distance, and by implication, against the actin-to-myosin centre-to-centre distance [43, 112].

This observation has important consequences for the sliding filament theory, because the distance the heads of the myosin molecules have to stretch out to reach the neighbouring actin must increase quite considerably as the muscle shortens. Within the range of lengths shown in the diagram, this distance varies by at least 70 Å from one extreme to the other (very long to fairly short). Since the heads themselves appear to extend out only about 135 Å from the centre of the filament in resting muscle they must be far more flexible than apppears on electron-microscope photographs, in order to be able to stretch the required maximum distance. It is probably the flexible neck region of the molecule (figure 1.2), which enables them to stretch in this way.

Another most important finding from X-ray studies is that the length of the primary filaments themselves does not appear to vary during contraction of a living muscle. It is possible to photograph a contracting frog muscle repeatedly with a suitably arranged low-angle X-ray set-up, where the camera shutter is synchronized to open only when the muscle contracts [44, 83]. The important X-ray reflections, which appear either during rest or contraction, do so at or near the meridian. They are the first and third orders, at 143 and 429 Å, of a basic myosin periodicity of 429 Å, and various orders of a basic actin periodicity of 360–370 Å. The 429 Å spacing represents the distance between heads along any one row of a myosin filament, and the 143 Å spacing the distance between a head in one row and its nearest neighbour in the next row. The actin periodicity at about 360 Å is the distance between cross-over points of the double-stranded actin helices, discussed earlier (cf. figure 1.1*a*).

None of the actin reflections alters significantly when the muscle contracts, showing that the length of the actin filaments remains unchanged during contraction, as predicted from the sliding filament model. Of the myosin reflections, the 143 Å spacing may increase in length by about 1 per cent showing that the myosin filament 'gives' slightly under the stress imposed on it. The *intensity* of this spacing decreases, however, to about 66 per cent of the rest value, and this, together with dramatic changes in the off-meridional reflections which themselves arise from the helical ordering of the cross-bridges, shows that considerable movement of myosin heads occurs during contraction. Most of this movement seems to be azimuthal, that is, a myosin head can swivel so that it points towards an actin filament 60° away from the one it points at in resting muscle. Reference to figure 1.1*b* will help the reader to appreciate how this movement can be effected.

Even more dramatic changes in intensity of the myosin layer lines occur during the onset of rigor, showing that myosin heads can then attach themselves to actin filaments at numerous points: the helical geometry becomes so disordered when they do so that the 429 Å spacing disappears entirely, and is replaced by layer lines which can be indexed either on a 360 Å or 380 Å helical repeat. On the other hand, the 143 Å spacing remains constant. Thus the backbone of the myosin filament remains unaltered during contraction or rigor, whereas the heads can swivel around to find suitable actin partners, outside the range to be expected on the rigid geometry of figure 1.1*b* and *c*. Nevertheless, during contraction, relatively few myosin heads seem to be attached to actin at any one time, and certainly far fewer than during rigor, presumably because on statistical grounds there is unlikely to be time enough for a head to find a suitable partner while rapid movement is going on. An interesting point about the X-ray diagram is that the myosin filaments in resting muscle are arranged with their longitudinal repeats in tranverse register across most of the width of each fibril. This accurate register becomes distorted during contraction and rigor, but most noticeably during the latter [83].

Changes in position of the myosin heads are clearly seen in insect flight muscle; during rest the heads are aligned at right angles to the filament axis, but during rigor change to an 'arrowhead' formation, pointing towards the Z-disc [130].

Great interest attaches to the X-ray studies, particularly to the most recent and more exact ones [83], because they represent the only really successful case, where X-ray analysis has yielded significantly new information about an intact living system. It is unfortunately impossible to discuss many other fascinating details of them here, but the interested reader is referred to the original papers, particularly to [44] and [83].

Table 1.1. Amino-acid composition of the structural proteins and their molecular weights [97, 122, 146, 147]
(Residues per 1000 residues)

Type of amino-acid		*Myosin* (1)	*LMM. Fr.I*† (2)	*HMM*† (3)	*Tropo-myosin* (4)	*Actin* (5)
Acidic	Aspartic	98·5	98·4	96·9	107	93·5
	Glutamic	182	249	162	254	109
	Free acidic groups‡	(174)	(219)	(153)	(284)	(114)
Basic	Lysine	107	112	102	128	50·2
	Histidine	18·5	24·9	16·5	6·6	20·1
	Arginine	49·8	71·2	40·2	49·1	48·6
Polar-un-charged	Threonine	51	39·1	52	31·1	75·4
	Serine	45·2	40·3	46	47·9	64·4
	Tyrosine	23·2	10·7	24·8	18·0	42·9
	Cysteine —SH	10·2	4·7	8·7	7·8	11·9
	Amide	(107)	(128)	(106)	(77)	(89)
Non-polar	Glycine	46·4	21·3	59·1	14·4	74·6
	Alanine	90·4	96·1	86·3	129	80·7
	Valine	49·8	45·1	56·7	32·3	49·7
	Isoleucine	48·7	46·3	52	35·9	73·2
	Leucine	93·9	114	86·3	114	68·8
	Phenylalanine	33·6	4·7	42·5	4·1	31·3
	Tryptophane	~5·0	—	—	—	10·2
	Methionine	26·6	22·5	30·7	19·1	43·5
Odd-man-out	Proline	25·5	0	37·8	2·0	50·2
MW × 10^{-3}		490–500	120–162	320–360	56	42*
% Helix		58	~100	45–51	95	8

† LMM = light meromyosin
HMM = heavy meromyosin
‡ Free acidic groups = (glutamic) + (aspartic) – (amide)
* Scopes, R. K. and Penny, I. F. (1970), in press.

2: Biochemistry of the Contractile Proteins

Myosin as an enzyme

We have so far discussed the way in which the molecules of myosin are organized within the thick filaments of the sarcomere, where their long, highly charged tails serve, by overlapping, to bind the filaments together, leaving the heads of the molecules protruding in orderly rows. It is the movements of these heads, either in a kind of sweeping motion or more probably undergoing miniature contraction cycles of their own, which are responsible for pulling or pushing the actin filaments towards the centre of the sarcomeres during a living contraction. The energy for this movement is derived from the free energy change ($-\Delta F$) when ATP is split at the active enzyme sites on the heads [31, 46].

By studying myosin on its own, we gather information on the nature of the splitting reaction and of the enzyme site, in the absence of any contractile activity. We must, however, be cautious in interpreting this information because the presence of the other partner in the contractile process, actin, vastly modifies the reaction mechanism, particularly the effects of metal ions.

The splitting reaction is common to a number of other nucleoside triphosphates (NTP's) besides ATP: the structure formulae of these is shown in figure 2.1. Designating myosin as M, nucleoside diphosphate as NDP, and inorganic phosphate as P_i, we can formulate the reaction as follows:

$$\underset{(1)}{M + NTP \rightleftarrows M{-}NTP} \overset{H_2O}{\underset{(2)}{\rightleftarrows}} M\Big\langle{}^{NDP}_{P_i} \underset{(3)}{\rightleftarrows} M + NDP + P \qquad 2.1$$

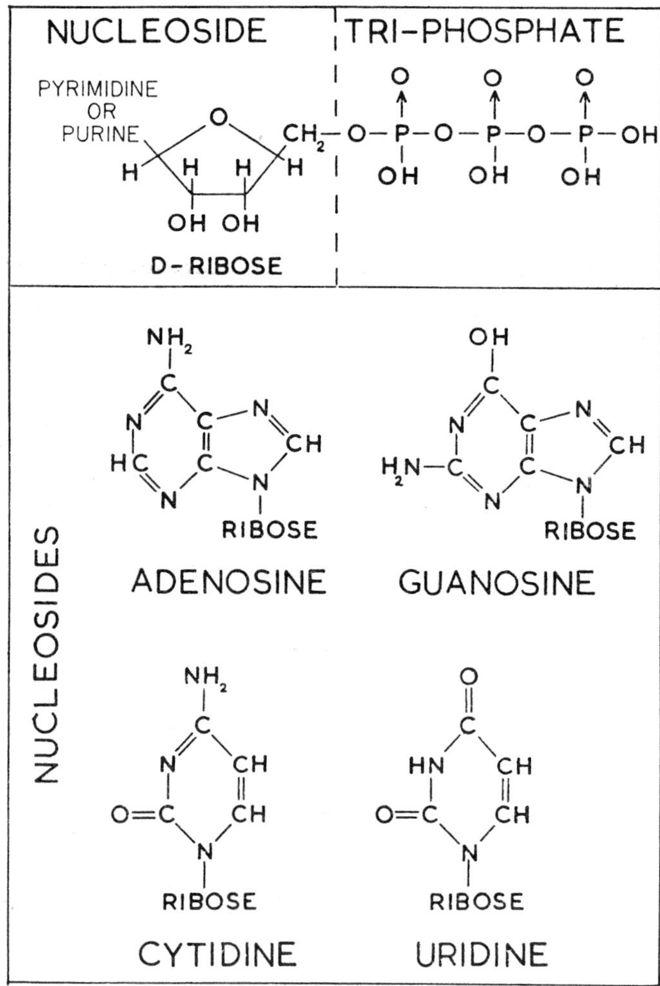

Figure 2.1. Conventional structural formulae for the various nucleoside triphosphates (NTP's), encountered in muscle. All except ATP are present in small or trace quantities. Note that inosine triphosphate (ITP) is not shown. It is the deaminated derivative of ATP with an OH-group in the 6-position of the purine ring.

A number of other steps can, of course, be written into this reaction scheme, particularly those involving adsorption of substrate or products on to the enzyme sites. We note at this stage that the reaction is quite strongly inhibited by the products, but more by NDP than by P_i [10, 116]. This inhibitory effect does not mean, however, that step 2 can be reversed to any considerable extent, because this step has a high and negative free energy change ($-\Delta F = 6$ to 10 kcal/mole), showing that the equilibrium is strongly to the right [116, 125, 169].

The phosphate moiety of the NTP tends to be converted more and more completely into the charged form shown on the left of equation 2.2, as the pH is increased towards 7·4, whereas the charge on the products tends towards the situation on the right. At the physiological pH of about 7·4, the total number of negative charges therefore increases by one as splitting proceeds, and hence it follows that a proton must be released during the reaction to equalize the change in charge, so that the pH falls [53]. We can schematize the changes at physiological pH as follows:

$$-\underset{\displaystyle O}{\overset{\displaystyle O^-}{\overset{|}{\underset{|}{P}}}}-O-\underset{\displaystyle O}{\overset{\displaystyle O^-}{\overset{|}{\underset{|}{P}}}}-O-\underset{\displaystyle O}{\overset{\displaystyle O^-}{\overset{|}{\underset{|}{P}}}}-O^- + H-\overset{*}{O}-H \rightleftarrows \tag{2.2}$$

$$-\underset{\displaystyle O}{\overset{\displaystyle O^-}{\overset{|}{\underset{|}{P}}}}-O-\underset{\displaystyle O}{\overset{\displaystyle O^-}{\overset{|}{\underset{|}{P}}}}-O^- + H\overset{*}{O}-\underset{\displaystyle O}{\overset{\displaystyle O^-}{\overset{|}{\underset{|}{P}}}}-O^- + H^+$$

(The 'starred' oxygen of water appears in the inorganic phosphate, P_i, formed) [98].)

At lower pH values than physiological, the negative charge on the inorganic phosphate decreases progressively, so that fewer protons are released during splitting and the acidification becomes less pronounced. This effect must always be taken into account in the measurement of enthalpy, or internal energy, changes because some of the latter will arise from the enthalpy of neutralization of the released protons (cf. Chapter 9).

Equations 2.1 and 2.2 are too naïve to represent the real state of affairs even during the splitting at ATP by myosin, because this reaction is strongly influenced by the ionic milieu, in two senses. First, it is extremely sensitive to the nature of the divalent cations which may be present and secondly, to the total ionic strength. In the complete absence of divalent cations, particularly Ca^{++}, it is doubtful whether the reaction can proceed at all, as we see on the left of figure 2.2*a*; this figure shows the effect of increasing the total Ca ion concentration at a constant ATP concentration of 2·5 mM, and an ionic strength ($\Gamma/2$) of about 0·1, mostly made up by K, Cl and tris-buffer ions. The maximal rate of splitting occurs when the Ca ion concentration equals or just exceeds the ATP concentration. Increasing $[Ca^{++}]$ to four times $[ATP^{-4}]$ has only a small inhibitory effect, and even increasing it to forty-fold reduces the rate only to about half maximal. On the other hand, there is a much faster and more drastic decline in rate as the Ca^{++} is reduced towards zero (results taken from [147], but see also [117]).

There are many ways of interpreting the effect of Ca^{++}, but it is important to bear in mind that ATP can chelate this ion fairly strongly, the binding constant† being about $2 \times 10^3\ M^{-1}$ [154]. Hence on the right of the peak in figure 2.2*a*, most of the ATP is in the form $CaATP^{-2}$, whereas on the left, increasing amounts of ATP^{-4} will be present. Furthermore, myosin itself has a strong affinity for both ATP and Ca^{++}, so that it is not immediately obvious what the substrate for the enzymic site really is. A simple explanation is that there is one site on myosin which will bind ATP^{-4}, but not the Ca complex, and another site close by, which will bind Ca^{++} only; then it is the interaction of the bound ATP with the bound Ca which causes splitting to occur [117]. Too much free ATP will tend to remove Ca^{++} from its site (left-hand side of the peak), and too much Ca^{++} will reduce the concentration of ATP^{-4} below that needed to saturate the ATP-site (right-hand side of the peak). This is tantamount to saying that the binding constant for ATP on to myosin is very much greater than that for ATP on to Ca^{++}, because of the slope of the curve on the right of figure 2.2*a*. Indeed, if we take the Michaelis constant ($K_M \cong$ dissociation

† By binding constant ($\overline{K}$) is meant the reciprocal of the dissociation constant (K). For a simple reaction $A + B \rightleftharpoons AB$, it is equivalent to [AB]/[A] [B].

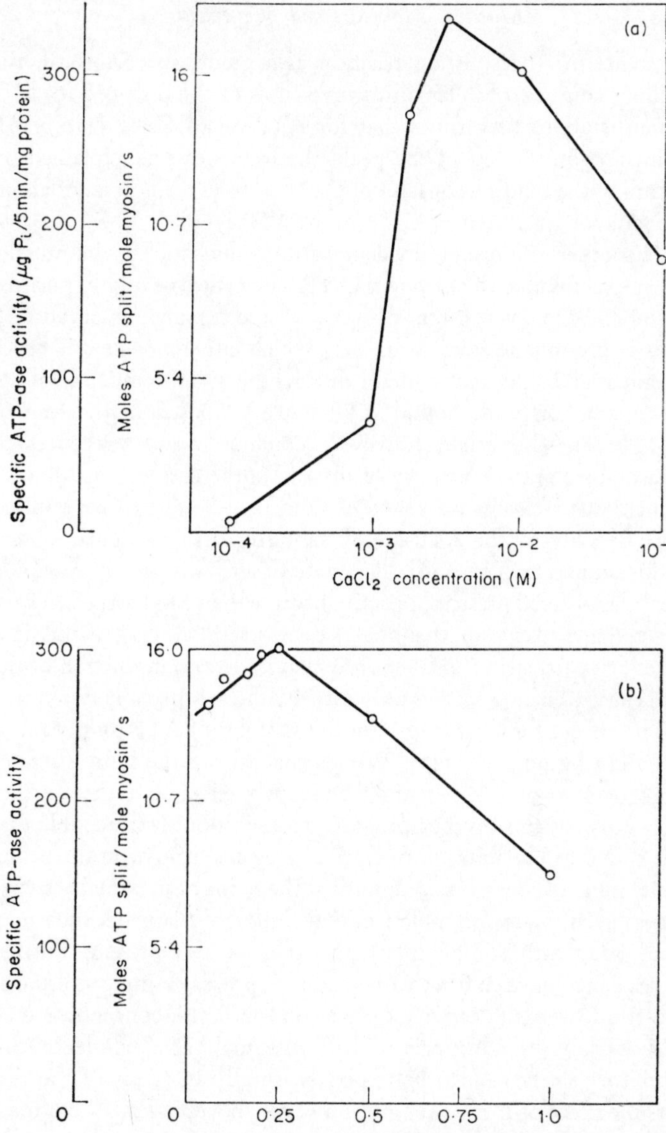

Figure 2.2. (a) Plot of the specific ATP-ase activity of myosin against the concentration of activating Ca ions (log scale). Tris-buffer pH 7·6; 0·05 M KCl; [ATP] = 2·5 × 10^{-3} M; temperature 25°C [148].

(*b*) Similar plot of the ATP-ase activity of myosin, but this time against the KCl concentration in the medium (= approx. ionic strength). Here the [Ca^{++}] was about 4 × 10^{-3} M. Other conditions as above [148].

constant) for the splitting reaction as a guide, its reciprocal, the binding constant of ATP on to myosin, is of the order of 10^5 M^{-1} which is about fifty times that for ATP on to Ca^{++} [116, 117]. Similarly, on the left of the peak, the results can be explained by assuming the binding constant of Ca^{++} on to myosin to be of about the same order as that of Ca^{++} on to ATP [169].

The other important divalent cation which occurs in muscle, Mg^{++}, seems able to activate the ATP-ase centre of highly purified *myosin* only to a very slight degree [147 and personal observations]. This is presumably due to its size, which either prevents it being bound at all to the active site or distorts this site too much to allow the interaction with bound ATP^{-4} we proposed in the case of Ca^{++}†. The literature, however, contains many references to measurable ATP-ase activity of myosin in the presence of Mg ions alone, but this is always very low, and may be due to contamination, possibly by the ATP-ase of the sarcoplasmic reticular system or by actin [138, 147, 149]. This state of affairs is greatly modified by the additional presence of actin, because then Mg has an *activating* effect, particularly in the presence of traces of Ca^{++}, and is a necessary partner in the reaction which leads to transduction of the free energy change into useful work by the actomyosin system.

The effect of ionic strength on the splitting of ATP by myosin is shown in figure 2.2*b* [147]. We see that raising the ionic strength from 0·05 to 0·25 has a slight activating effect, but that further increase leads to slow but progressive inhibition. It is possible that this is due to breaking up of large aggregates of myosin molecules as the ionic strength is raised, until at the point of maximal activity, the molecules are assembled into the shorter filaments shown in plate I.2*a*, with some dimers and trimers also present. Further increase in ionic strength, we may suppose, produces single molecules, and this lowers the activity once again, until at a concentration of 1 M KCl, it has dropped to half maximal. The whole of this inhibitory effect cannot be explained simply in terms of the dispersing effect of ionic strength, however: beyond $\Gamma/2 = 0{\cdot}6$ most of the myosin already exists in monomer form, as shown by recent electron microscope and ultracentrifugal studies [93, 94], and hence

† For an alternative explanation, see [168].

the increasing inhibitory effect of ionic strength beyond this point must evidently be due to altering the local distribution of charge in the region of the active site. Incidentally, the EM and ultracentrifugal studies we have referred to provide definite proof that the formation of regular, thick filaments of myosin by side-to-side aggregation of monomers occurs only in the physiological conditions of about pH 7·4 and an ionic strength of about 0·15; considered purely as a myosin ATP-ase, these more complex filaments are clearly not as effective as the somewhat shorter and thinner filaments, found at $\Gamma/2 = 0{\cdot}25$ [81, 93, 94].

There are many other interesting effects of ions which could be mentioned here, for example the effects of Zn, Mn, Co, Sr, Ba and NH_4^+; and the surprising activating effect of the divalent ion chelating agent, ethylene-diamine-tetra-acetate (EDTA), particularly in the presence of NH_4^+, which together can increase the ATP-ase activity of myosin well beyond what can be achieved with Ca^{++} alone [146]†. These effects are, however, of little direct importance in a discussion of muscle contraction, though they provide useful clues to the mechanism of splitting and the nature of the active site. More vital are the questions of the role of the thiol-groups (SH) of myosin, of the nature of the NTP and of the effect of pH on the NTP-ase activity.

The role of the SH-groups of myosin in the splitting reaction has been elucidated rather elegantly by the use of a new reagent [147] which is rather more specific in its effect than those used previously, such as *p*-chloro-mercuri-benzoate and N-ethyl-maleimide [122, 149]. This symmetrical reagent, 5-5′-dithiobis (2-nitro-benzoic acid) or DTNB, reacts with SH-groups by itself splitting in half; the extent of the reaction is measured by the amount of coloured product formed. Of the 35 to 40 SH-groups in the myosin molecule DTNB will react with 9 or 10; the others are evidently sterically hindered from reacting with such a large molecule. 50 per cent of the ATP-ase activity of myosin is lost when 5 groups are bound, and inactivation is complete when 7 are bound. These groups are almost certainly in the head region of the molecule, probably fairly close to the active site [147].

† Also see footnote opposite.

The real point of interest in these SH-studies lies, however, in the difference in numbers of groups bound in the absence or presence of $MgATP^{-2}$, which can attach itself to the active site of myosin without being split, as we have seen. By adding $MgATP^{-2}$ first, 2 SH-groups per 250,000 particle weight become unavailable to DTNB; these are evidently those directly involved in the binding of ATP or its divalent metal complexes to the active sites (on each of the two heads of myosin), possibly through the purine part of the ATP molecule by hydrogen-bond formation (cf. also similar, but earlier findings in [142*a*]). These findings also underline the importance of keeping the SH-groups of myosin intact during its preparation by including a reducing agent, such as mercapto-ethanol, at each stage to prevent oxidation [122].

The nature of the NTP is decisive in determining the rate at which it will be split by myosin [61, 146]. NTP's with an OH-group in the 6 position of the purine ring (ITP, UTP and GTP) are split faster than those with an NH_2 in that position (ATP, CTP). The OH and NH_2 groups can form hydrogen bonds, and are almost certainly involved in the attachment of substrate. Like a number of other features of the reaction the order of rates at which the various NTP's are split is completely reversed when actin is present as well [61]. The reason for this is not known at present, though the fact that three point attachment of substrate probably becomes necessary when actin is also a partner in the reaction, would mean that each of the ionic or hydrogen bonds involved would be altered in their relative strengths.

The effect of pH on the rate of splitting of the various NTP's by myosin has been studied in some detail [6, 139]. Here we need only consider the special case when the splitting of ATP is catalysed by Ca ions. The pH activity curve is shown in figure 2.3, curve a. The activity approaches two maxima, one of which lies below pH 6·0 and the other above pH 9·5, whereas the broad minimum of activity lies, strangely enough, in the physiological range of pH from about 7 to 7·5. There is no obvious explanation of these maxima and minima in terms of the charged groups on the protein, although the decline in activity from pH 6·0 to pH 7·5 might be attributed to removal of a proton (H^+) from the imidazole group of the histidyl residues (pK 6·5 to 7·0). The fairly sharp increase in activity above pH 8·0 might likewise be attributed to the gradual removal of a proton from the phenoxyl groups of the tyrosyl residues or the terminal amino groups of the lysyl residues (pK 9·5 to 10·0).

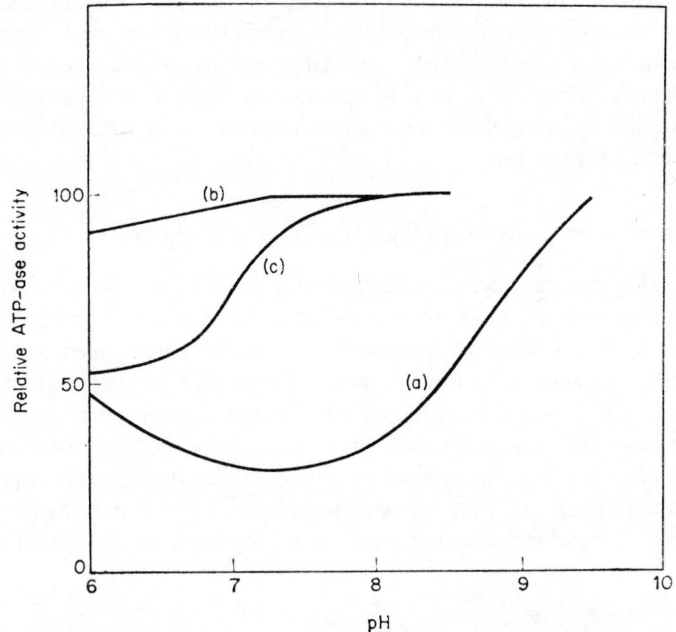

Figure 2.3. Plot of the relative ATP-ase activity against the pH of the medium: Curve (a) for myosin in 0·25 M KCl + 5·0 × 10^{-3} M ATP and Ca^{++};
(*b*) for myofibrils at an ionic strength of 0·12, activated by Ca^{++} only (5 × 10^{-3} M). [ATP] = [Ca^{++}];
(*c*) for myofibrils activated by Mg^{++} (5 × 10^{-3} M) + Ca^{++} (1 × 10^{-4} M). [ATP] = 5 × 10^{-3} M.
(*a*) taken from [6] and [139] (*b*) and (*c*) author's observations.

The pH/activity curve is vastly modified when actin is also present, as shown for the particular case of myofibrils in figure 2.3, curves b and c. Curve b, where Ca^{++} is the sole activating ion, is almost independent of pH from 6·0 to 8·3. On the other hand, when Mg^{++} is the dominating divalent cation, and only traces of Ca ions are present, the activity rises on an S-shaped curve from a minimum at pH 6 to a maximum at 8·5, the midpoint lying at about pH 7·0 (curve c). It is tempting to attribute this also to the changing charge on the histidyl imidazole groups, because the curve closely resembles their proton dissociation curve in shape. Here though, actin has so modified the effect that it operates in the reverse sense to that with myosin alone, because now removal of the proton from these groups at slightly alkaline pH results in activation instead of inhibition. There is probably another activity maximum in the region of pH 9·3 (not shown). It is, of course, this Mg/Ca-activated myofibrillar system which most closely resembles the real situation during contraction.

Modification of myosin ATP-ase by actin

The reactions between myosin and actin which are of direct interest to our subject take place when actin is in the F- or fibrillar form, that is when it consists of the long double-stranded superhelices of actin monomers, shown in figure 1.1*a*. When a colloidal solution of such F-actin is added to myosin at high ionic strength ($\Gamma/2 = >0{\cdot}3$) the immediate effect is to increase the viscosity of the solution greatly, due to formation of complex aggregates of actomyosin [144, 145, 146]. These aggregates are somewhat of the form of the arrowheaded structures of the F-actin H-mero-myosin poly-

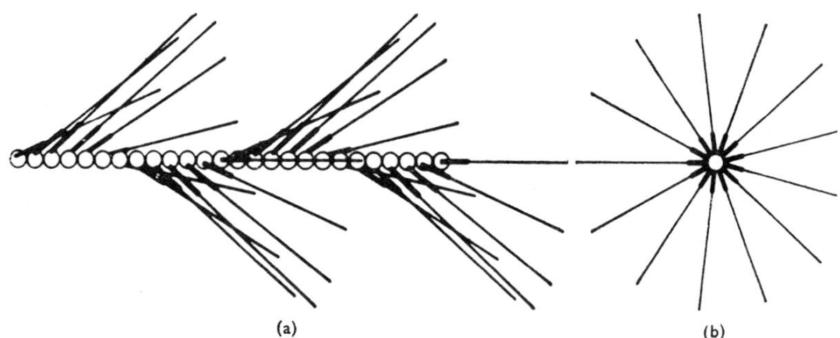

Figure 2.4. Possible structure of part of an actomyosin 'molecule' in solution, at the optimal myosin/actin ratio of 4/1 by weight. One long double-headed myosin molecule is attached between each pair of actin beads, to give an arrowheaded, spiral structure, such as that seen in plate 1.3*b*. Dimensions only approximate. (Drawing by Mr Stewart Brown, Meat Research Institute, Langford, Bristol.)

mers, shown in plate I.3*b*; however, intact myosin molecules have much longer tails than those of H-meromyosin, and this evidently prevents their attachment, one to every actin monomer as in the case of H-meromyosin, so that the *maximum* molar combining ratio of myosin to actin now becomes 1 to 2, instead of 1 to 1. We can envisage the resulting, complex polymers of F-actomyosin to be of the form shown in figure 2.4; they are of indefinite length and thus of very variable molecular weight. The polymer shown in the figure would have a molecular weight (MW) of about 16 million, so that it is not at all surprising that 'solutions' of actomyosin have a high viscosity. Such polymers often aggregate together to form complex ladder-like structures [133], probably through union of their tails.

Reactions in 'solution' at high ionic strength

ATP reacts with solutions of purified actomyosin at $\Gamma/2 = 0{\cdot}6$ in different ways according to the nature of the divalent cation present. On addition of $MgATP^{-2}$, there is a sudden and dramatic decline in the viscosity of the solution, which falls back to the much lower additive viscosity of the two components, and remains at this level almost indefinitely [8, 143, 144, 145, 146]. The polymers have been completely broken up into their constituents by the $MgATP^{-2}$, and this occurs by the latter attaching itself at the active enzymic site on myosin, and repelling the attached actin electrostatically [127]. Since $MgATP^{-2}$ is not an effective substrate for myosin ATP-ase, it is not split to any significant extent during this process [147].

Replacement of $MgATP^{-2}$ by $CaATP^{-2}$ results in a much slower and more incomplete decline in the viscosity [8], and this is accompanied by vigorous splitting of ATP, because $CaATP^{-2}$ is an excellent substrate for myosin (figure 2.2*a*). Hence it does not take long for the ATP to be used up, and then the viscosity returns immediately to the high value characteristic of actomyosin polymers [8, 146]. Thus $MgATP^{-2}$ is a much better 'plasticizer' of actomyosin than the Ca chelate, in spite of the fact that the latter would also be expected to have a considerable electrostatic repelling effect. It should be noted that ATP can only act as a plasticizer in this way when it is in the form of a divalent cation chelate, the univalent cations alone being almost totally ineffective [8].

Gel system at physiological ionic strength

Although the effects we have described take place under the very unphysiological conditions of high ionic strength, they have one feature in common with actomyosin *gels* and with the more organized system of intact myofibrils, at the physiological ionic strength of ~0·15, that is: *MgATP^{-2} acts as a plasticizer of actomyosin complexes and breaks them up into their constitutents, when the conditions are such that it cannot itself be split at the active site.* Unfortunately it is here that we run into a confusing feature of the gel system at low ionic strength, which has only recently been fully recognized: its reaction towards MgATP^{-2} depends upon the way in which it is prepared.

Gels of the actomyosin directly extracted from muscle, and also of artificial actomyosin made from actin accidentally contaminated with 'tropomyosin', both behave in the 'orthodox' manner towards MgATP^{-2}, that is to say, the gel will neither super-precipitate nor split ATP in the presence of Mg ions alone [36, 37, 153]. Even the traces of Ca ions contained in the reagents and the glassware (~10^{-5} M) are, however, sufficient to set the rapid splitting of ATP in train, and this is accompanied by the dramatic super-precipitation of the gel which was so elegantly demonstrated in the early studies of the contractility of actomyosin [144, 145]. To obtain the full plasticizing effect of MgATP^{-2} with natural actomyosin gels and to inhibit the ATP-ase, it is necessary to chelate out all the contaminating Ca ions, preferably with EGTA which binds Ca^{++} 1000 times more effectively than it does Mg^{++} [154]. Looked at the other way round, Ca^{++} is necessary for superprecipitation and splitting to occur in the presence of MgATP^{-2}, but the natural actomyosin system is so sensitive to this ion that half the maximal rate of splitting is already obtained at the very low free $[Ca]^{++}$ of 3×10^{-7} M [36, 37, 153, 154].

Can we also say that Mg^{++} is necessary for super-precipitation?

Indeed we can, because high concentrations of $CaATP^{-2}$ *alone* give only an incomplete effect, although the splitting of ATP still occurs at a rapid rate [123, 153, 154]; addition of a trace of Mg ions now produces immediate and complete super-precipitation. We can summarize these effects with natural actomyosin as follows: (i) natural actomyosin gels, contaminated with 'tropomyosin', are 'plasticized' by $MgATP^{-2}$ and will not split it or super-precipitate; (ii) addition of Ca^{++} in trace amounts catalyses splitting and super-precipitation in the presence of Mg^{++}; (iii) even in massive amounts Ca^{++}, in the absence of Mg^{++}, will not catalyse super-precipitation but only splitting; (iv) for super-precipitation to occur Mg ions are necessary as well as Ca ions.

We now turn to the properties of *artificial* actomyosin, prepared from purified actin and myosin. Gels made from these two proteins differ in one essential respect from those prepared from natural actomyosin: they do not require the presence of traces of Ca^{++} to be able to super-precipitate and split ATP in the presence of Mg^{++}; in other words, they have lost their Ca sensitivity [38, 115]. They will, however, split $CaATP^{-2}$ alone just as well as natural actomyosin does, and like it, they do not then super-precipitate. The essential difference therefore lies solely in their reaction to $MgATP^{-2}$ which no longer inhibits splitting of ATP, and hence cannot catalyse plasticization. Although arguments are still in progress about the nature of the Ca-sensitizing factor missing from pure actomyosin [122], it seems very likely that it is tropomyosin plus troponin [20, 38, 115]. It has indeed been shown that such 'tropomyosin' restores Ca sensitivity to actomyosin if it is prepared in the presence of a thiol-compound or other reducing agent, to keep the SH-groups of the proteins in their reduced state [20, 115]. As we showed in the last chapter, tropomyosin itself is probably also involved in the Z-disc structure from which it may run together with troponin in either direction in the groove between the double helices of the actin filaments.

One possible hypothesis which would explain the action of 'tropomyosin' is that its presence within the helices of 'natural' actin repels the $MgATP^{-2}$ or ATP^{-4} which is attached to the myosin partner, and thus prevents the interaction leading to splitting and the onset of contraction. When Ca^{++} is added, it would then bind to negatively charged groups on the troponin part of the tropomyosin complex, neutralize the charge, and thus allow interaction to occur between the actin monomers and the substrate/myosin complex [35*a*]. In other words, Ca^{++} works not as an activator in the strict sense, but as an inhibitor of an inhibition (via the charge effect on tropomyosin). This indeed is to go back to a hypothesis proposed in the very early days of the discovery of the Marsh factor, but through a completely new mechanism [13, p. 251]. Moreover it reinstates $MgATP^{-2}$ or ATP^{-4} as the true substrate for actin/myosin ATP-ase [10], as opposed to $CaATP^{-2}$ which is split by myosin without involving the actin partner (cf. figure 2.2*a*). This hypothesis resembles an idea put forward in [122] (figure 12), but with 'tropomyosin' replacing the hypothetical ESG factor, postulated there. A necessary rider to this concept is, however, that 'tropomyosin' should not be able to bind Mg^{++}, but only Ca^{++}. This important point has recently been settled by direct experiment [35*a*].

Before leaving artificial actomyosin, and turning to the more orderly array of actin and myosin filaments in myofibrils, we should repeat that the various NTP's we mentioned in connection with pure myosin now have exactly the reverse effect, that is to say, NTP's with an NH_2 group in the 6-position of the purine or pyrimidine ring (ATP, CTP) are split by actomyosin more rapidly than those with an OH group in that position (ITP, UTP, GTP) [8]. This shows, as we said, that the substituent in the 6-position of the purine ring is one of the partners in the essential links between actin and myosin during the contractile process, the other being the triphosphate grouping.

ATP-ase activity and contraction of myofibrils

Myofibrils may be prepared for the study of their ATP-ase activity and contractile properties by homogenizing muscle in weak buffers at pH 7·0 and an ionic strength of 0·1 to 0·15. With rabbit muscle, the best fibrils are obtained by allowing the muscle to go into rigor before homogenization (about 20 hours after death, at 10°C):

these have identical ATP-ase activity to those prepared from pre-rigor muscle. The fibrils can be separated from connective tissue and intact fibres by differential centrifugation of the homogenate at low speed [123], or more simply by filtration through butter muslin [10]. They are then washed thoroughly to remove soluble contaminating proteins. Such fibrils are 1 to 2 μ in diameter and 10 to 100 μ in length.

This system has considerable advantages over actomyosin gels, because not only are the actin and myosin filaments more highly organized, in their natural array and proportions, but also Ca-sensitivity is preserved for some days. Hence the observations can be more readily related to contraction *in vivo*.

The degree of contraction of the fibrillar system may be measured in relative terms by centrifuging the fibrils at about 3000 $\times$ g, and then measuring the volume of the fibril layer, before and after the addition of ATP [109]. By making use of a simple stroboscopic effect it is also possible to watch the fibril layer as it is centrifuged down, so that rates of sedimentation can be obtained (see figure 2.5*a*). These rates can then be related to the rate of splitting of ATP or other NTP's under like conditions. This method gives a measure of the average particle size of the fibrils *after* they have contracted; it is virtually impossible to catch them during the contractile phase itself, because this is so rapid that it takes place during the initial acceleration of the centrifuge to full speed. There is also the anomaly that the highest rates of sedimentation are given by contracted fibrils, because they are denser, whereas a low rate indicates that the fibrils were swollen and had contracted little if at all before centrifugation began. This effect is due to the pushing out of osmotically held water during contraction.

The fibrillar system closely resembles natural actomyosin gels in its reaction toward divalent ions. In the presence of $MgATP^{-2}$ only, with EGTA present to chelate out all traces of Ca ions, it splits ATP at less than a twentieth of the maximal rate in the presence of traces of Ca^{++} (see figure 2.5*b*). Under these conditions, the fibrils do not contract, but on the contrary tend to swell, as we see from the rates and extent of sedimentation shown in figure 2.5*a* (curve A). On the other hand, when Ca^{++} at a concentration of about 0·4 mM is

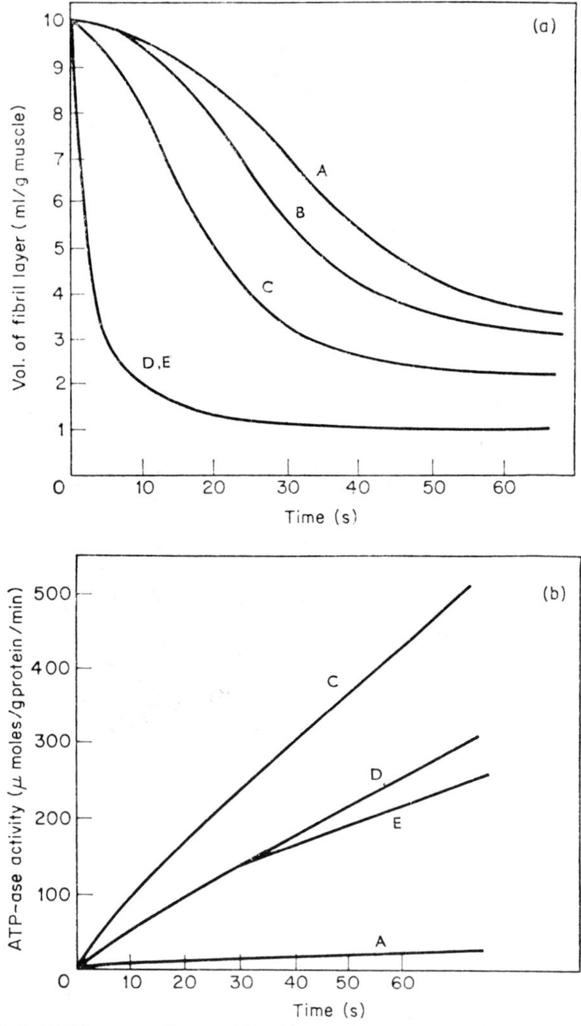

Figure 2.5. (*a*) Effect on volume of fibril layer of treating fibrils in various media and then centrifuging them at 3000 × g (time plot). Imidazole buffer pH 7·2; ionic strength = 0·12; [ATP], where added, = 4×10^{-3} M; temp. = 20°C.

(A) ATP + 4×10^{-3} M $MgCl_2$ + ditto EGTA;
(B) Control, no ATP;
(C) ATP as in A, but *no* $MgCl_2$. [$CaCl_2$] = 4×10^{-3} M;
(D) ATP as in A, 4×10^{-3} M $MgCl_2$ + $0{\cdot}4 \times 10^{-3}$ M $CaCl_2$;
(E) ATP as in A, $0{\cdot}4 \times 10^{-3}$ M $MgCl_2$ + 4×10^{-3} M $CaCl_2$.

(*b*) ATP activities of fibrils under the various conditions of (*a*) above. (Author's observations.)

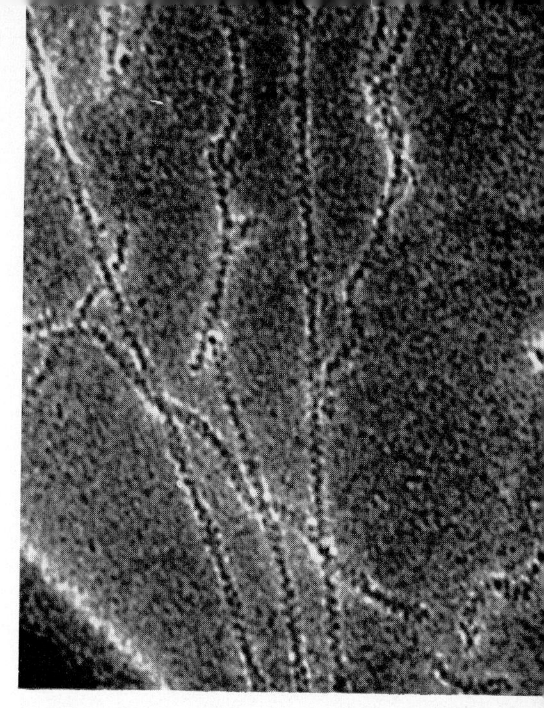

te 1.1c Purified F-actin preparation, ectly from muscle from which myo- had been previously extracted. te the double helical structure. h of the small beads or monomers actin are about 55 Å in diameter. (Photo by courtesy of Dr H. E. Huxley.)

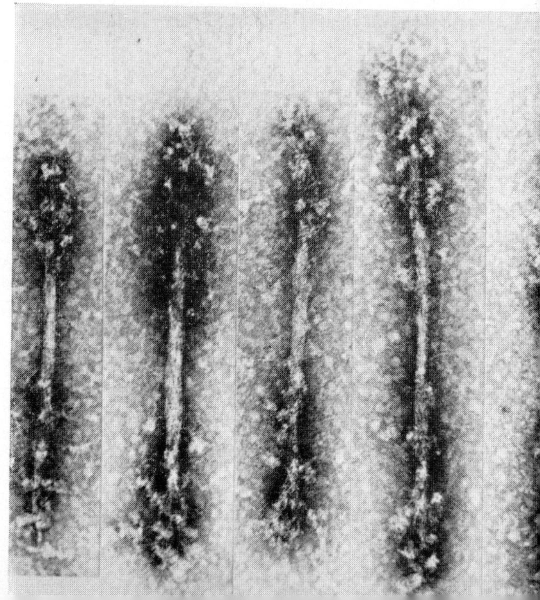

1.2 Artificially prepared thick nents, made by lowering the ionic ngth of a myosin solution from to 0·15. The filaments range in th up to about 1·4 μ. Note the sin heads sticking out sideways, pt in the bare central region, re there is only overlap of myosin . This region is about 1700 Å long. (Photo by courtesy of Dr H. E. Huxley.)

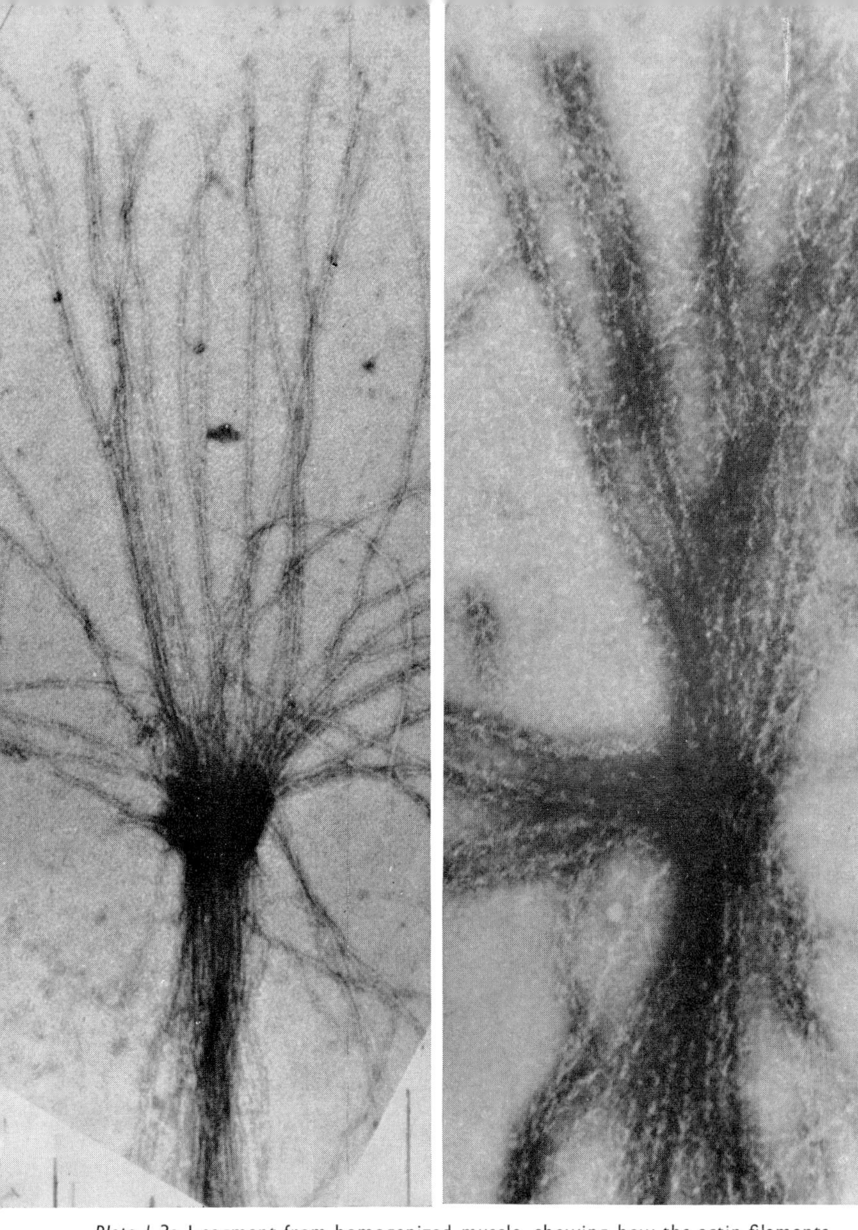

Plate 1.3a I-segment from homogenized muscle, showing how the actin filaments of the I-band are attached to Z-disc.

Plate 1.3b Effect of treating an isolated I-segment, as above, with a solution of H-meromyosin. Note the formation of arrowheaded structures, where the H-meromyosin has attached itself, one molecule to every actin monomer. Also note the different sense of the arrowheads on each side of the Z-disc.

(Photo by courtesy of Dr H. E. Huxley.)

Plate 1.4a EM photo to show detailed structure of a sarcomere. Note the thin actin filaments of the I-band, and how they run between the myosin filaments of the A-band, up to the edges of the H-zone. Also note the heads of the myosin molecules protruding from the thick filaments, except in the centre of the sarcomere, near the thickened M-line.

Plate 1.4b Enlargement of the H-zone region of a sarcomere to show more detail of myosin heads, and of the M-line.

100 Å

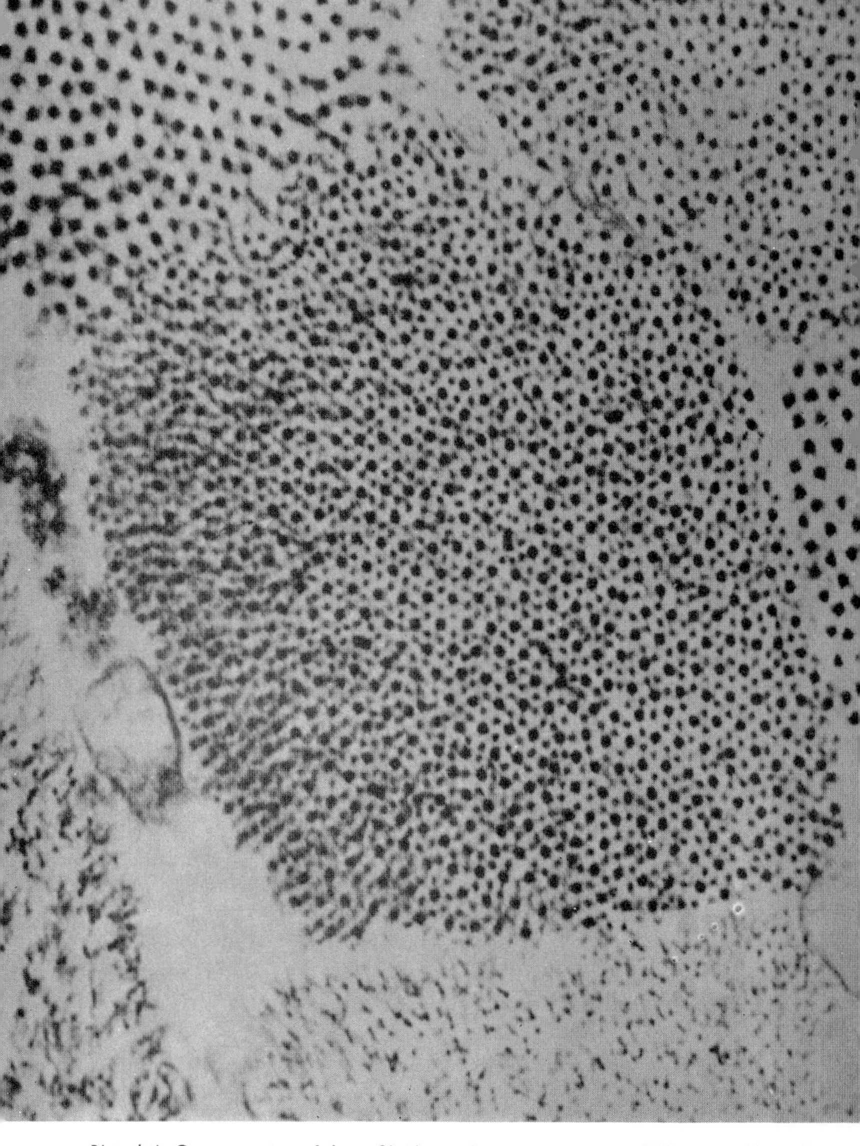

Plate 1.4c Cross-section of three fibrils, to show arrangement of filaments. Note the hexagonal arrangement of actin filaments around the myosin filaments in the fibril in the centre. At the bottom, part of a fibril is shown cut through the I-band, where only actin filaments can be seen, and at the top left is another fibril cut through the M-line region, where the union of myosin tails gives a triangular shape to the filaments.

(Photo by courtesy of Dr H. E. Huxley.)

added to fibrils in the presence of 4 mM $MgATP^{-2}$, it leads to their rapid and extensive contraction, as shown by the high rate and extent of sedimentation, accompanied in this case by *maximal* ATP-ase activity (figures 2.5*a* and *b*, curves D) (see also [154]).

Compared with its effect, even at very low concentration, in the presence of $MgATP^{-2}$, the Ca ion *alone* has a far less pronounced effect on the rate of sedimentation, as we see from the figures, where it was added as $CaATP^{-2}$ at 4 mM concentration (curve C). Nevertheless the ATP-ase activity (curve C) is even higher than that of the Mg system in the presence of traces of Ca ions (curve D). It is probable that if all traces of Mg ions had been removed before addition of $CaATP^{-2}$, the rate of sedimentation would have been even lower, but this is not easy to achieve in practice.

Thus, like the actomyosin gel system, fibrils react to ATP in different ways according to the divalent cation present: in the presence of Mg^{++} only, they split ATP very slowly and do not contract, but rather tend to swell: addition of a trace of Ca ions (0·01 to 0·4 mM) as well leads to rapid splitting and massive and complete contraction: on the other hand, Ca^{++} alone, even at high concentration, has little or no contractile effect, although it allows a very high rate of splitting, all of the energy from which must be wasted as heat. Finally, addition of Mg ions (0·4 mM) to the Ca system, although it somewhat reduces the splitting rate, also *immediately* catalyses full contractile activity (cf. figures 2.5*a* and *b*, curves E).

These observations give us an excellent insight into how the contraction-relaxation cycle might operate in living muscle. In the relaxed state, $MgATP^{-2}$ only would be present, all traces of Ca ions having been chelated out in some as yet unspecified manner. Then when the stimulus for contraction arrived, Ca ions would be released to a free concentration of about 10^{-5} M ($\equiv$ 0·01 mM) and in a total amount somewhat greater than that needed to saturate all the active sites on the heads of the myosin molecules or elsewhere (between 0·1 to 0·25 μg ions of Ca^{++} per g of muscle). Rapid splitting of ATP and contraction would then set in immediately. For the muscle to relax, the postulated Ca-chelating system would need to begin to operate at full capacity once more, and at the same time

ATP would either have been fully resynthesized from phosphocreatine, or have become available again by free diffusion to the active sites from the sarcoplasm. The reformation of $MgATP^{-2}$ would then result in the inhibition of splitting and complete relaxation. As we shall see, just such a system does in fact exist in the intact muscle in the form of a Ca^{++} pump, situated in the tubules of the sarcoplasmic reticulum which enwraps each fibril within a fibre.

More about the Ca^{++} effect with fibrils

The quantitative aspects of the Ca-activation of the fibrillar ATP-ase and of contraction are of importance to a discussion of contraction *in vivo*, because they indicate the theoretical requirements to be demanded of the Ca-pump we mentioned above, and they furnish a clue to the nature of the reaction at the active sites of myosin during the natural process. They also present another interesting anomaly, as we shall see.

Typical results for the Ca-effect with rabbit myofibrils are given in figure 2.6*a*, cf. [154]. In this figure, the conditions are close to those which actually obtain in living muscle, e.g. a temperature of 35°C, an ionic strength a little below 0·15, a pH of ~7·2, and an initial concentration of $MgATP^{-2}$ of 4 mM. The free Ca^{++} concentrations were controlled in this experiment by means of Ca-EGTA buffers which act similarly to the more usual pH buffers, except that it is the concentration of free Ca^{++} which they control [154]. These values are plotted on the abscissa as $\log_{10}$ $[Ca^{++}]_{free}$ and the relative ATP-ase activity is plotted on the ordinate.

It should be noted that the absolute maximal value of the splitting rate, with sufficient Ca^{++} present ($\geq 10^{-5}$ M) is very consistent from sample to sample of rabbit myofibrils under these conditions, and lies close to 2100 μ atoms P_i split off per minute per g of protein. This represents an approximate maximal turnover rate of 36 moles ATP split per mole of myosin per sec, or one mole split per mole myosin in 28 millisec (assuming that the fibrillar protein contains ~50 per cent myosin which represents a concentration of $\sim 10^{-4}$ M in intact, fresh muscle).

With *fresh fibrils* the Ca-activation curve is sharply S-shaped; the splitting rate falls off very rapidly at Ca concentrations below 10^{-6} M, and the half value is reached at $10^{-6.5}$ M Ca^{++}. Even at $[Ca]^{++} = 10^{-8}$ M, however, there is still a small residual non-Ca activated ATP-ase activity, amounting to between 5 and 10 per cent of the maximum. The origin of this activity is still unclear; it is far higher than that required for the energy production in resting living muscle, which is below 0·1 per cent of the maximal rate during a contraction, but it has been persistently reported under different conditions by other workers [154]. It could possibly be controlled in intact muscle by a 'soluble' factor, reported to inhibit the Mg-activated splitting of ATP by artificial or desensitized actomyosin [122].

Interest attaches to the residual non-Ca-activated splitting because it can be increased greatly by storing fibrils under aseptic conditions for some days, in the presence of air or oxygen. Curve 2 in figure 2.6*a* illustrates the effect of storing fibrils in oxygen for 6 days at 20°C, where we see that the non-Ca activated splitting at $[Ca^{++}] = 10^{-8}$ M increased from about 8 per cent to 52 per cent of the maximum rate, although the absolute value of the latter remained virtually unchanged. In parallel with their much higher non-Ca-activated ATP-ase activity, fibrils stored in this way contract nearly maximally in the presence of $MgATP^{-2}$ alone, as shown by the centrifuge test we have described. This is the corollary of the superprecipitation of *purified* actomyosin in $MgATP^{-2}$.

The loss of Ca-sensitivity on storage of fibrils is indeed very reminiscent of the state of affairs with purified actomyosin [38]: there the lost Ca-sensitivity could be restored, as we saw, by adding 'tropomyosin' with its SH-groups protected during preparation by a thiol reagent, but not by *adding* 'tropomyosin' prepared in the usual way in the presence of air [33, 115]. Acting on this hint, fibrils were stored under nitrogen with and without addition of 5 mM mercapto-ethanol, for comparison with those stored under oxygen. The results are shown in figure 2.7, for the splitting rates of the various preparations at $[Ca^{++}] = < 10^{-8}$ M (i.e. almost zero Ca^{++}). We notice that the low initial splitting rate, characteristic of these conditions, gradually increases on storage in O_2, but is completely

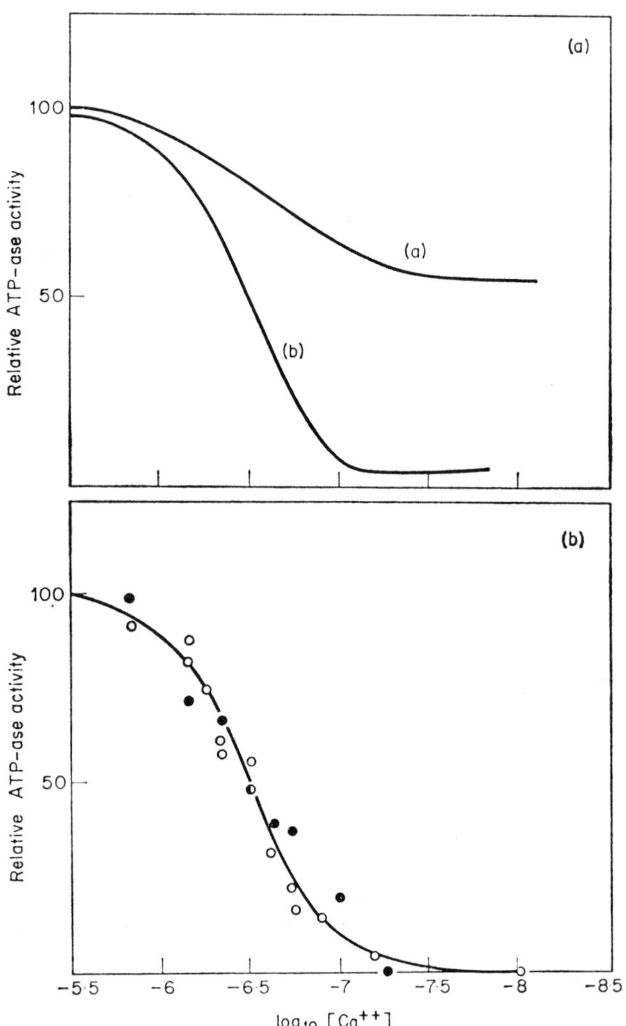

Figure 2.6. (a) Effect of [Ca^{++}] in the medium on the relative ATP-ase activity (V/V_m) of myofibrils, in the presence of 5×10^{-3} M ATP/$MgCl_2$. Conditions as figure 2.5, except that temp. = 35°C. [Ca^{++}] controlled by the use of Ca/EGTA buffers, and given as $\log_{10}$ [Ca^{++}]. The bottom curve is for fresh fibrils, and the upper curve for fibrils, stored under aseptic conditions in oxygen at 20°C for six days.

(*b*) The results of (*a*) replotted after correction for the residual ATP-ase activity (V_r) found at a vanishingly small [Ca^{++}] of 10^{-8} M. The ordinate term now = $V - V_r/V_m - V_r$. Circles for fresh fibrils, and dots for fibrils stored in O_2 for six days at 20°C. (Author's observations.)

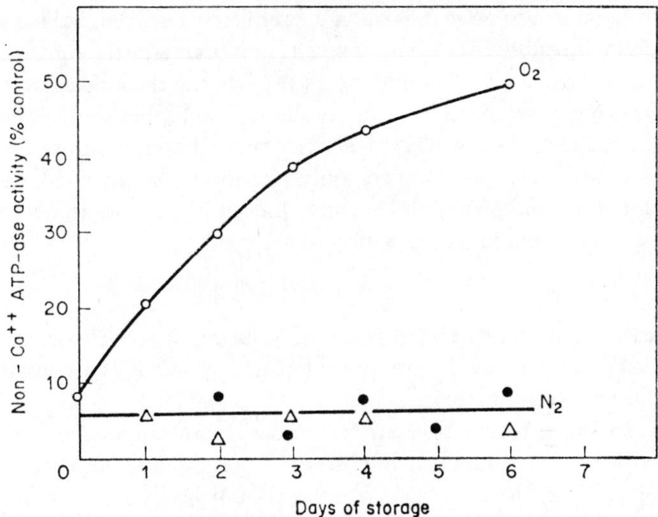

Figure 2.7. Plot of the residual ATP-ase activity of fibrils, measured as in figure 2.6, against days of storage at 20°C. Note that the concentration of contaminating Ca ions was kept below 10^{-8} M by adding 4mM EGTA. Results given as % control ATP-ase activity in presence of 1×10^{-4} M Ca^{++}.

o storage in oxygen;
● storage in nitrogen;
△ storage in $N_2 + 5 \times 10^{-3}$ M mercapto-ethanol.
(Author's observations).

unaffected in the presence of N_2, with or without 5 mM mercapto-ethanol. Hence we deduce: (i) that oxidation of SH-groups is the most likely cause of the loss of Ca sensitivity; (ii) that these SH-groups are probably not situated near the active site of myosin itself, because the maximal rate of splitting of $MgATP^{-2}$ in presence of adequate Ca^{++} is unaffected by storage; (iii) that the SH-groups involved probably belong to the 'tropomyosin' which we postulated to lie within the actin double helices, and to be the keystone of the Ca-sensitivity.

How many sites for Ca^{++} at the active centre?

The question of the number of Ca ions involved per enzyme site, when the fibrillar system is in its fresh and fully Ca-sensitive state with Mg ions also present, is still not completely resolved. Exchange studies with radioactive Ca have given equivocal results, somewhere between 1 and 2 Ca^{++} per site [154, 158], whereas the kinetics of the process quite clearly and unequivocally demand 2 per site, because of the shape of the log $[Ca^{++}]$/activity curve shown in figure 2.6*a*. This curve is far too sharp for only one ion to be involved, and certainly not sharp enough for three. Indeed, the results accurately obey a simplified kinetic equation of the type:

$$V'_m/V_x = 1 + K/[Ca^{++}]_x^2 \qquad 2.4$$

where V'_m is the corrected maximal velocity, V_x is the corrected velocity at a Ca^{++} concentration of $[Ca^{++}]_x$, and K is a constant (see legend to figure 2.6).

This simple form of equation necessarily pre-supposes that the enzyme site is saturated with substrate: this is the case here because the $[ATPMg^{-2}]$ was 4 mM, whereas the Michaelis constant (K_m) which determines the degree of saturation is below 8×10^{-5} M; hence the term $K_m/[ATP]$ of the Michaelis–Menten equation is less than 0·02 and can be ignored [10]. The constant in equation 2.4 has a numerical value of 10^{-13} M^2, under the conditions of figure 2.6*a*, from which it follows that the $[Ca^{++}]$ necessary for half the maximal velocity of splitting is $10^{-6 \cdot 5}$ M. The justification for this equation is given in figure 2.6*b*, where the activity (V/V_m) has been corrected for the residual non-Ca^{++} activated splitting, and then plotted against log $[Ca^{++}]$. The circles are for curve a in figure 2.6*a*, where fresh fibrils were used and the non-Ca^{++} activity was very low, whereas the dots represent the results after 6 days storage in O_2 at 20°C, where the non-Ca activity had risen to 52 per cent of the maximal. Both sets of points lie close to the theoretical line which was calculated from equation 2.4. The results given in [154] fit a similar curve, but with a different mid-point of about $10^{-6 \cdot 3}$ M Ca^{++}, probably because the temperature of the experiments was 25°C, instead of the 37°C used here.

Hence, in the case of Ca activation of fresh fibrils in the presence of $MgATP^{-2}$, it is fairly certain that two Ca ions are somehow involved at or near the active centre, whereas for the simple Ca-activated splitting of ATP by *myosin* itself, or by fibrils in the *absence* of Mg^{++}, there is little doubt that only one Ca ion is necessary [117]. It is, of course, the Mg/Ca-fibrillar system which much more closely resembles the actual state of affairs in muscle, where the true substrate is either ATP^{-4} or $MgATP^{-2}$, and Ca^{++} probably acts by binding to unspecified negatively charged groups of 'tropomyosin' within the actin helices, thus neutralizing their inhibitory effect. In that case, the effect of oxidizing the SH-groups of this 'tropomyosin' is easily explained because it would clearly result in a very drastic change of shape of the molecules, and prevent them from any longer fulfilling their anti-ATP-ase role. This hypothesis would also explain the otherwise puzzling fact that the maximal ATP-ase activity of the fibrillar system stays numerically constant, regardless of whether tropomyosin has been oxidized or not, that is, whether the system needs Ca ions or only Mg^{++}. This constancy would be expected if $MgATP^{-2}$ or ATP^{-4} is the true substrate for the actin/myosin interaction, but not so if $CaATP^{-2}$ were the substrate when 'tropomyosin' was in the reduced state, but was suddenly and mysteriously replaced by $MgATP^{-2}$, as soon as this protein complex was oxidized.

We have seen earlier that it is indeed most likely to be troponin which binds the Ca^{++} required for 'activation' of myofibrils; because of the molar ratios of the proteins, this would give an apparent binding of 2 Ca^{++} per myosin molecule, which might explain the form of equation 2.4 (cf. [35*a*]). Note also that only Ca^{++} is bound, not Mg^{++}.

The effect of temperature

The effect of temperature on the myofibrillar ATP-ase activity must be taken into account if this activity is to be related to contraction *in vivo*, particularly when considering the contractile activity of the muscles of cold-blooded animals, which have been the subject of so many exact studies of the energetics of contraction [11, 67].

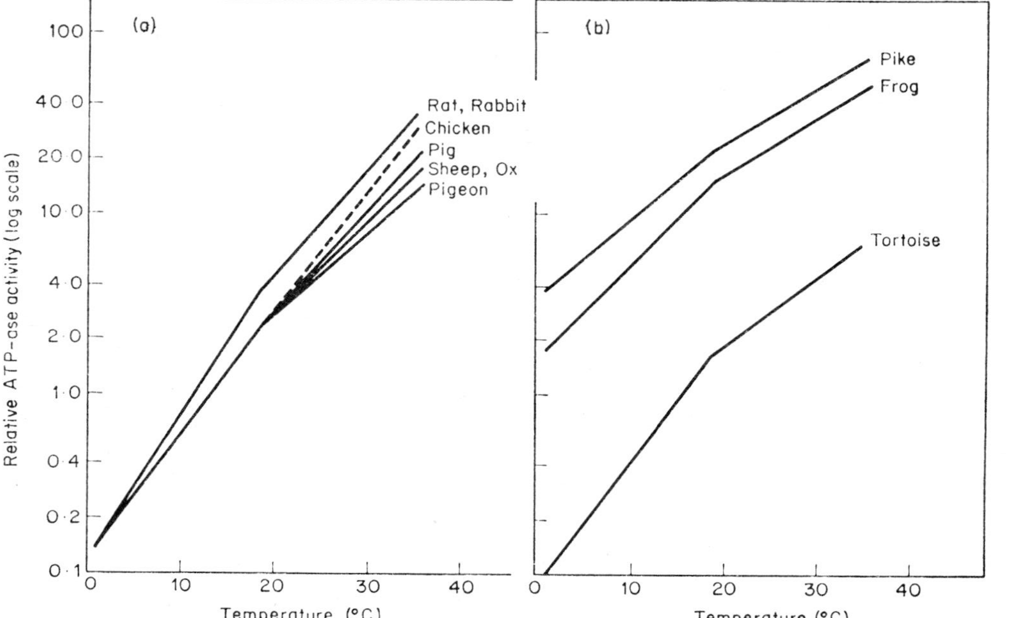

Figure **2.8.** (*a*) Relative ATP-ase activities (log scale) of various species of homothermic animals plotted against temperature (°C).

(*b*) Ditto for heterothermic animals.

Muscles used: mixed leg muscles of rat, frog and tortoise; longissimus dorsi of rabbit, pig, sheep and ox; breast muscles of chicken and pigeon.

$[ATP] = [Mg^{++}] = 4 \times 10^{-3}$ M; $\log_{10} [Ca^{++}] = 3·7$; pH = 7·2 to 7·6; imidazole buffer (cf. [11]).

For our present purpose it is sufficient to consider the effect of temperature under the following standard conditions:

$[MgATP^{-2}] = 4$ mM; $[Ca^{++}]_{free} = 0{\cdot}4$ mM; imidazole buffer = 40 mM; pH = 7·2 to 7·6; ionic strength adjusted with KCl to 0·15 for mammals and birds, and to 0·12 for amphibia and fish.

The results for nine species of animal are given in figures 2.8*a* and *b*, where the myofibrillar ATP-ase activity is plotted on a logarithmic scale against the temperature in °C (see also table 2).

We see from the figure that the slope of the plot is much steeper for the homothermic mammals and birds than for the heterothermic pike, frog and tortoise, that is, the activity of the first group of the animals is far more affected by temperature than that of the second. There are also interesting differences within the groups: for instance, the muscles of the large mammals split ATP at their body temperature of about 37°C at only half the rate shown by the white muscles of the rat and rabbit at this temperature, whereas of the two birds, chicken (breast) closely resembles rat and rabbit, and pigeon (breast) the larger mammals. At first sight, these results seem to indicate that 'white' muscles such as leg muscles of the rat, back muscles of the rabbit and breast muscles of the chicken are fast, and that the redder muscles of the large mammals and of the pigeon's breast are slow. But this association of redness with slowness is really too naïve, since the slowest muscles of all, those of the tortoise, are also the whitest of all and contain almost no myoglobin, the pigment which accounts for the redness of the muscles of the higher animals.

It is safer to regard the differing ATP-ase activities of the various species as adaptations to function: indeed it is quite probable that a limited number of types of myosin (iso-myosins), differing in their specific ATP-ase activities, were evolved amongst the lower organisms, and that from these, two main groups were selected out, one by the heterothermic animals and the other by the homothermic; the latter are characterized by being more affected by changing temperature than the former [122]. All the animals within a group would then possess the genetic code necessary for producing a variety of iso-myosins, each adapted to a specific muscle function, within the limits of the group. To make the point clear, let us consider the growing rabbit. Born immature, this animal possesses at birth only a 'slow' type of myosin, of about the same activity as that of the adult pigeon breast [122, 148]. In most of its fast 'white' muscles, however, the rabbit rapidly develops a type of myosin with at least double this activity, whereas in one of the few slow 'red' muscles it possesses, the semitendinosus, the ATP-ase activity remains at or below the immature level; in fibrils from the adult rabbit semitendinosus, for instance, the ATP-ase activity is less than one third of that of the white longissimus dorsi or psoas muscles (cf. also [139]).

An objection to the above argument is that we have only considered the myosin partner in the actomyosin complex, and that actin might also play a role in determining the overall fibrillar ATP-ase activity. This is unlikely to be so for two reasons: first, the actins from a wide variety of species will form actomyosin complexes with rabbit myosin which are almost indistinguishable in their ATP-ase activities; second, the relative splitting rates of the myosins, isolated from various species, agree well with those of the corresponding fibrillar preparations, where actin is also present [11, 122, 139]. Hence, it is evidently the groupings close to the active centre of myosin which differ from species to species and from iso-myosin to iso-myosin within a species, whereas neither the light meromyosin tails of the myosins nor their actin partners need have been modified very markedly, at least amongst the vertebrates. This generalization is not necessarily valid for the specialized myosins found in smooth muscles, nor those in lower animals, such as worms and molluscs [140].

To put the effects of temperature into quantitative terms, it is necessary to consider the energies or more correctly, the enthalpies of activation which are given in table 2. Also shown are the entropies and free energies of activation. These parameters have been calculated from the Eyring equation for absolute reaction rates, discussed in [51].

Briefly, the enthalpy of activation represents the extra internal energy the reacting molecules must acquire, by collision, before the reaction can take place: that is, it is the amount of energy required to pass the 'energy barrier'. In the complex case of muscle contraction, at least four 'molecules' are involved: a myosin head, a molecule each of ATP and Ca^{++} and a monomer in the actin filament. The activation energy is thus distributed among several degrees of freedom, because a relatively highly organized complex must be formed before the reaction can take place. The formation of such a complex represents an increase in order, that is a decrease in entropy. For this reason, the entropy of activation might generally be expected to be negative, which is exactly opposite to what is found amongst the examples quoted in the upper half of table 2.1.

The free energy of activation is derived from the other two functions, and can also be calculated independently. It is this function which determines the rate of reaction at a given temperature. From the table it is seen to be reasonably constant, whatever type of fibrillar preparation is studied, whether 'warm' or 'cold-blooded', and for this reason it will not be discussed further here.

The most significant differences in table 2 occur between the enthalpies and entropies of activation of the homothermic and heterothermic groups respectively, whereas both functions are reasonably constant with groups. On average, the enthalpy of activation of the homothermic group is almost double that of the heterothermic, and the entropy of activation is high and positive for the former and low or negative for the latter.

The tortoise, in the temperature range from 0 to 18·5°C, is a possible exception to the rule, because the enthalpy approaches that of the homothermic group and the entropy is also, like that group, positive instead of negative. This may arise from the difficulties of measuring the rates of splitting ($= K'$ in the table) sufficiently accurately at 0°C, where they are very low indeed. Nevertheless, two other observations on tortoise muscle agree with those given here: (i) the decay of active tension after a single muscle contraction in living tortoise muscle has an apparent enthalpy of activation of 19 k.cal/mole and an entropy of +23 units within this temperature range [22]; (ii) a comparison of the rate of heat liberation during a tetanus at 0°C, between tortoise and frog muscle, shows that the latter liberates heat 12 times faster than the former [72, 73]. The rate of heat liberation particularly depends on the rate of the underlying chemical reaction, the splitting of ATP, so that it may be compared directly with the relative rates of splitting by the respective fibrillar preparations in table 2: there again the ratio of the frog to the tortoise rates is also about 12. Hence it seems we can rely on the values for tortoise at 0°C, so that it is a genuine exception to the heterothermic rule.

Further discussion of the meaning of the positive and negative entropies of activation found in the table is not very fruitful at present, because we do not know enough about the exact mechanism of the splitting of ATP to be able to say why there should be a difference between the groups (but see [22, 51]). It is relevant, however, to note a peculiarity of the fibrillar preparations, also shown by pure actomyosin and myosin [116]: all three ATP-ase systems, when activated by Ca ions in massive concentrations in the absence of Mg ions, show lower enthalpies of activation than any in the table; the average value is about 10 kcal/mole, independent of temperature. Moreover, the entropy of activation becomes −11 units. The essential difference between this type of ionic system and the fibrillar system in its physiological state with Mg ions and a trace of Ca ions present, is that neither actomyosin nor fibrils contract in the presence of Ca ions alone. Hence it is evidently the transduction of free energy into movement and work which modifies the enthalpies and entropies of activation, through the intervention of actin in one case and not in the other. In fact, if we suppose as we did earlier, that Ca ions activate the system in the presence of $MgATP^{-2}$ by removing the inhibition imposed by charged groups of 'tropomyosin' protruding from within the actin helices, then there is one more step to consider from the point of view of enthalpy and entropy of activation. This may well explain the differences in these parameters between the simple myosin-$CaATP^{-2}$ system and the much more complex actin/'tropomyosin'/myosin system of natural fibrils and living muscle.

Table 2. The enthalpy (ΔH), entropy (ΔS) and free energy (F)Δ of activation of the myofibrillar ATP-ase of various species. (Author's observations; cf. [11].) The rate constant K', is given as moles ATP split/mole myosin/second.

Animal	Temp. °C	K′ (≡ velocity of splitting)	ΔH k.cals per mole	ΔS Entropy units	ΔF k.cals per mole
Homotherms					
Rat, rabbit	0·6	0·14			
			28·2	+39·4	17·1
	18·5	3·80			
			23·4	+24·1	16·2
	35·0	35·30			
Bullock, pig,	0·6	0·14			
sheep			23·6	+23·5	17·0
	18·5	2·20			
			21·6	+16·6	16·6
	35·0	16·70			
Chicken	0·6	0·13			
			29·0	+39·0	17·0
	18·5	2·50			
			23·0	+20·0	16·8
	35·0	29·60			
Pigeon	0·6	0·27			
			23·0	+23·0	17·0
	18·5	2·30			
			20·0	+16·0	16·5
	35·0	14·20			
Heterotheams					
Tortoise	0·6	0·10			
			25·4	+29·0	17·2
	18·5	1·70			
			14·4	−9·8	17·3
	35·0	6·70			
Frog	0·6	1·30			
			17·0	+4·8	15·8
	1·5	16·70			
			14·0	−4·5	15·8
	35·0	44·00			
Pike	0·6	3·7			
			15·4	−0·9	15·7
	18·5	22·0			
			12·7	−9·2	15·5
	35·0	74·0			

For conditions, see p. 53.

3: Model Fibre Systems

In the last chapter we saw how isolated actomyosin and the fibrils prepared by homogenization of muscle, split ATP at physiological ionic strength, and how this splitting was accompanied under some conditions by synaeresis of the gel or contraction of the fibrils, whereas under others, little or no contraction occurred and the energy from the splitting process was wasted as heat. We also saw that if splitting was inhibited, particularly in the presence of Mg ions, ATP then acted as a plasticizer of the system, manifested as swelling of the gels or fibrils. In such a system, however, the contractile and relaxing effects of ATP can only be demonstrated qualitatively, because there is no easy way of loading the very short fibrils or the gels, in order to measure the work they do or the tension they develop during the contraction–relaxation cycle.

A closer approach to the conditions actually obtaining in muscle can be made by using so-called 'model' systems of fibres, or of actomyosin threads. Here we shall discuss the model fibres, prepared by treating bundles of fresh fibres with aqueous glycerol for some days at 0°C [45]. This system has great advantages over actomyosin threads, because the actin and myosin filaments are then preserved in their natural array, whereas they are poorly aligned in the artificial threads and do not develop high tensions on the addition of ATP [160]. In glycerolated fibres it is only the filaments of actin and myosin whose structure is preserved intact; other cell components of metabolic importance in the living muscle have been destroyed or removed; for example, most of the soluble glycolytic enzymes are leached out, because the semi-permeable membrane, or plasmalemma, of the fibres is partly destroyed by the glycerol treatment; similarly, the mitochondria lying between fibrils and the sarcoplasmic reticulum which enwraps each fibril are severely damaged and unable any longer to operate effectively. Glycerolated fibres can therefore be regarded solely as an organized contractile system of actin and myosin filaments permeable to divalent cations and to ATP and many other large anions.

The question of diffusion

Although the glycerolated fibre has become permeable to large ions, it still presents a barrier to the diffusion of metabolites such as ATP, for the following reasons: (i) ATP can diffuse into the fibre only at a rate equal to or less than its rate of diffusion in water; (ii) as ATP diffuses inwards it is split at the ATP-ase sites on the actin and myosin filaments of each successive fibril towards the centre of the fibre. Hence unless the rate of diffusion is at least equal to the maximal rate of splitting at these sites, the ATP concentration will tend to fall to zero at some depth within the fibre; for an ideal system, this depth can be determined by the Meyerhof-Schultz diffusion equation for cylindrical fibres:

$$r = (4c.D/A)^{\frac{1}{2}} \qquad 3.1$$

where r is the limiting thickness of the outer ring of 'active' fibrils; c is the external ATP concentration; D is the diffusion coefficient and A is the maximal rate of splitting of ATP by the outermost fibrils [111]. In other words, the limiting thickness depends on the square root of the ATP concentration, or, for a fibre or bundle of given thickness, the limiting ATP concentration depends on the square of the radius. The average radius of single glycerolated fibres is approximately 60 μ, which is about forty times that of the fibrils, discussed in the previous chapter.

Unfortunately ideal diffusion is far from applying to the early experiments on which we are forced to rely for most of our information on the development of tension or the performance of work by single fibres or fibre bundles. This is mainly because it was not realized until quite recently that the Ca ion played a major role in the ATP-ase activity and contraction of so-called Mg-activated systems of fibrils or fibres, so that in most of the experiments only Mg ions and ATP were added to the fibres; these reagents would, if they had been pure, have had no contractile effect, as we saw in the last chapter. However, it is clear now that the reagents used were always more or less heavily contaminated with Ca, so that there was usually sufficient of the latter present by accident to stimulate splitting and contraction fully [154]. It is useless to take the argument further than this, because we can only guess at the exact conditions which obtained. We shall therefore regard all of the earlier results as only semi-quantitative, and accept with some hesitation the calculations of the limiting thickness and of the diffusion coefficient of ATP, which were based upon them. Indeed, an exact value for the diffusion coefficient is still not available, although the available evidence suggests that it lies in the range 2 to 8×10^{-8} cm^2/sec ([60], but see [17] for a much higher value).

Development of tension

The development of tension by single glycerolated fibres can be measured by allowing them to contract against a strong torsion wire [155]. Under these conditions, the length of the fibres changes very little as they attempt to contract, so that the contraction is almost truly isometric. To measure the small movements of the torsion wire a light beam is directed on to a small mirror attached to it and the reflected spot of light focused on to a length scale or on to photographic paper on a revolving drum.

In figure 3.1, the tension developed by single rabbit psoas fibres on the addition of ATP is plotted against the total ATP concentration. In these experiments, the ionic strength is 0·1 to 0·15 and the Mg ion concentration 1 to 2×10^{-3} M. Hence, above total ATP concentrations of $\sim 1 \times 10^{-3}$M, the concentration of free ATP^{-4} increases. This free ATP progressively tends to chelate adventitious contaminating Ca ions and since these latter are necessary for contraction (see figure 2.5*a*) we should expect two opposing effects of increasing the ATP concentration: (i), ATP will become available to an increasing number of fibrils in depth, as its external concentration is raised, so that the overall splitting rate of the fibre will tend to increase and with it the tension developed; (ii), more and more Ca^{++} will be chelated out as the ATP concentration is raised above that of the Mg ions, with the result that a secondary inhibitory effect of ATP will follow in the wake of the first 'activating' effect

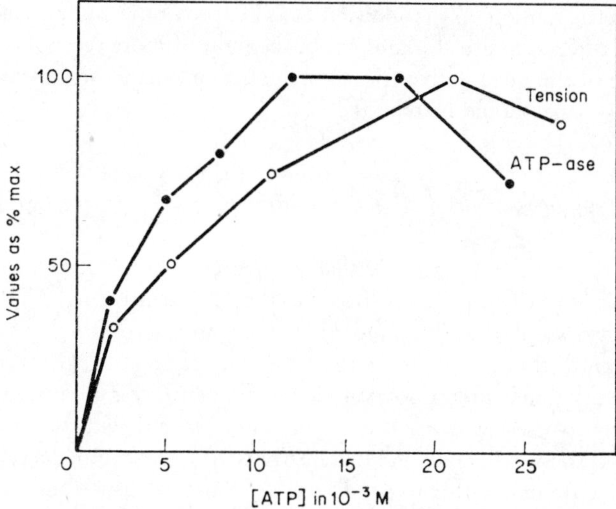

Figure 3.1. The effect of [ATP] on the tension developed by single glycerolated rabbit muscle fibres (after [155] and [159]). Included for comparison are the author's measurements of the ATP-ase activity of a preparation of large fibre fragments, containing very few fibrils.

$[Mg^{++}] = 1$ to 2×10^{-3} M; ionic strength = 0·10; temp. = 20°C. Contamination by Ca^{++} probably proportional to [ATP].

of its increasing availability. In the limit, the inhibitory effect, reinforced also by the increasing ionic strength due to ATP itself, will become dominant and the splitting rate will fall once more, and with it the tension developed. This is exactly what happens experimentally, when the overall ATP concentration begins to exceed 20×10^{-3} M, as we see from the figure.

Because of the two opposing effects we have mentioned, it is clear that these early experiments rarely, if ever, gave maximum values for the tension. Usually the value seems to have been in the range of 1 to 1·5 kg/cm² of fibre cross-section, although much higher values of about 4 kg/cm² are also quoted, where more contaminating Ca ions were probably included accidentally in the experiments [155]. The higher value for the tension is close to that for fully stimulated, living rabbit muscle. Thus, in the limit under ideal conditions, it seems that the glycerolated fibre will develop almost maximal isometric tension on the addition of ATP, providing Mg ions and a trace of Ca ions are also present. In the light of more recent knowledge of the quantitative effects of the Ca ion, many of these early experiments should be repeated.

ATP splitting and the development of tension

In the last chapter we saw that the degree of contraction or synaeresis of fibrils was closely linked to the Mg/Ca activated ATP-ase activity (see figures 2.5 *a* and *b*), but this in itself is no proof that the contraction derived its energy from the splitting of ATP, because the work done or tension developed by the fibrils was not measured, for comparison with it. We are on surer ground with the model fibre system, where the ATP-ase activity of fibre fragments can be measured under similar conditions to those used in the tension studies.

Values for the ATP-ase activity of fibre fragments are given by the dots in figure 3.1 (cf. also: [155, 159]). Here the ATP-ase activity reaches its maximum at considerably lower ATP concentrations than the tension, and the decline due to the over-optimal effect of ATP also sets in earlier. The reason for these differences between one system and the other lies in their different thicknesses, that is, in the differing magnitudes of the diffusion barrier to ATP (see equation 3.1). The fibre fragments, being shorter and somewhat thinner, become fully saturated with substrate at lower ATP concentrations than do the intact fibres, so that both the maximum of the splitting curves and the ensuing decline are reached earlier than the maximum and decline in the tension experiments. After allowance for this difference, the splitting of ATP correlates with the development of tension about as closely as we could expect. It is therefore certainly justifiable to conclude from these experiments that the energy for contractile activity comes directly from the splitting process under the specific ionic conditions we have described. An important corollary also holds good: that is, agents which reduce the ATP-ase activity of model fibres also reduce their capacity to develop tension or power.

Isotonic contraction and the development of power

Besides the tension model fibres can develop, when held at constant length (= isometric conditions), it is also important to know the power developed when they are allowed to shorten against a constant load (= isotonic conditions). For this purpose it is simpler to use a small bundle of fibres rather than a single one.

The first point to establish is the power (= rate of doing work) developed at varying ATP concentrations against a constant load: a typical experiment at two ATP concentrations is shown in figure 3.2, where time is plotted on the abscissa and work done on the ordinate. At both concentrations, there is a very short delay of about 0·6 sec, after addition of ATP to the bath, before the muscle begins to shorten and do work; this is followed during the first 2 sec by a rapid acceleration, not discernible on the scale used; thereafter the rate declines once more as the fibre bundle shortens further. Raising the ATP concentration from 2 to 11×10^{-3} M increases the initial rate of doing work about fourfold. In this experiment the dimensions of the bundle were about $300 \times 300\ \mu$ by about 1·5 cm long, and it contained about twenty fibres in its cross-section. The nominal load was 1000 g/cm^2 of bundle.

To calculate the maximal power developed during the course of doing work, it is sufficient for our purpose to measure the slopes of the work/time curves 10 sec after addition of ATP. The power is given by:

$$\text{Power} = \Delta W/\Delta t = -P\Delta l/(M \times 10 \text{ sec}) \qquad 3.2$$

where W is the work done in g.cm, P is the load on the bundle in g, $-\Delta l$ is the length change in cm, and M is the corrected weight of the bundle in g. Power is then given in g.cm/g.bundle/sec.

The collected results for a number of experiments at constant load, and at two different thicknesses of bundle, are shown in figure 3.3, where power is plotted against [ATP]. These findings should be compared with those in reference 17, where quite different

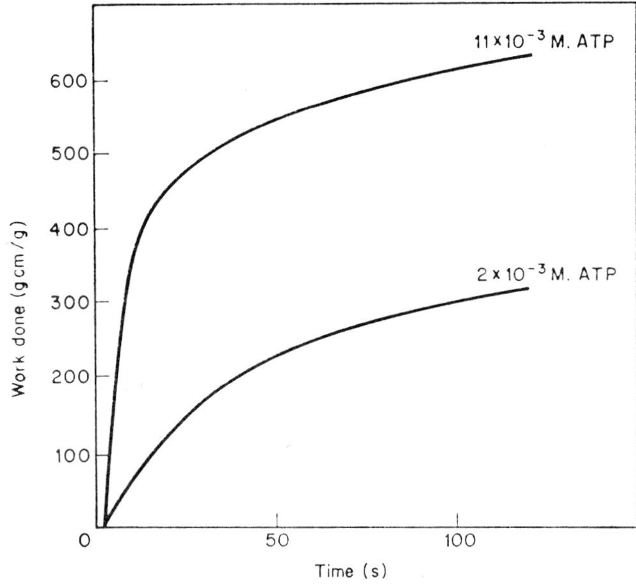

Figure 3.2. The effect of [ATP] on the time course of shortening of rabbit psoas fibre bundles (glycerolated). Results plotted as work done by the loaded bundles against time.

$[Mg^{++}] = 2 \times 10^{-3}$ M; ionic strength = 0·12; imidazole buffer, pH 7·2; temp. = 20°C.

Bundles 300 × 300 μ in cross-section, 1·5 cm long; load/cm² 1000 g; total shortening in 240 sec (*a*) 0·29 l_0 (*b*) 0·48 l_0; at 2×10^{-3} and 11×10^{-3} M. ATP, respectively. (Author's observations.)

results were obtained, which lead to almost diametrically opposite conclusions. The reader must take his pick.

There are several strange features of the power/[ATP] curves, not least the fact that although $[Mg^{++}]$ was only 2×10^{-3} M, yet the power continued to increase as $[ATP]_{total}$ was increased towards the very high level of 24×10^{-3} M. At this point the $[ATP]_{free}$ would have been about 22×10^{-3} M, easily sufficient, one would think, to chelate out all the contaminating, but essential, Ca^{++}, and thus to inhibit contraction [154]. However, this was obviously not the case, and we conclude, therefore, that contaminating Ca^{++} had been brought in with the ATP and increased proportionately as [ATP] was increased. This would satisfactorily

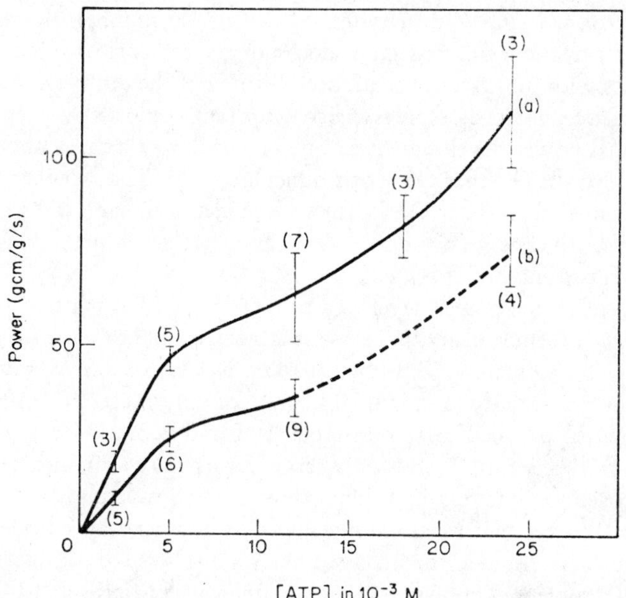

Figure 3.3. Power developed by fibre bundles at varying [ATP]. Conditions as in figure 3.2.

(*a*) Bundles 200 × 200 μ in cross-section; load/cm² = 1200 g.
(*b*) Bundles 300 × 300 μ in cross-section; load/cm² = 700 g.

Vertical bars = SE's of the means; numbers in brackets = number of experiments. (Author's observations.)

explain the absence of a power plateau, followed by a decline, expectable at the highest ATP concentrations.

Another anomalous feature is the comparatively rapid increase in power between [ATP] of 0 and 5×10^{-3} M. This probably arises first, because the number of actively contracting fibres in the bundle increases sharply in this concentration range, as equation 3.1

predicts, and secondly, because the total load on the bundle was held constant. Hence at low [ATP], when the proportion of active fibres would also be low, the actual load on these fibres would be excessive, gradually becoming more nearly optimal as [ATP] was increased and with it the proportion of actively contracting fibres. The difficulty of judging the optimal loading is, of course, one of the drawbacks of working with a system in which the number of active fibres per cm^2 of cross-section does not remain constant, but increases with substrate concentration, because of the diffusion effect. In practice, however, there is a fairly wide range of loads over which optimal power development can be obtained, so that it is only at very low [ATP] that the worst difficulties arise. The fact that the experimental power/[ATP] curves are of identical shape at the two different thicknesses of bundle, as equation 3.1 would predict, is at least reassuring in this respect.

To relate the power developed to the rate of ATP splitting puts us into a further quandary, since it is not possible to estimate the latter rate accurately with fibre bundles, because of diffusion difficulties, particularly in the initial stages: not only must ATP diffuse in, but the products to be measured, ADP and P_i, must diffuse out, and this necessarily introduces large errors. We can, however, approach the question in a different way by considering the theoretical power which can be developed at 100 per cent efficiency, when ATP is split at the maximal rate by fibrils, where diffusion is not a problem. The maximal rate of splitting by fibrils at 20°C is approximately equivalent to $0{\cdot}8 \times 10^{-6}$ moles/g.muscle/sec [10]. With a free energy of hydrolysis of ATP of $\sim$10 kcal/mole ($\equiv 427 \times 10^6$ g.cm of work/mole at 100 per cent efficiency), the maximal possible power development would thus be 342 g.cm/g.muscle/sec. The maximum observed from figure 3.3, for the thinner of the two bundles, is 112 g.cm/g/sec, giving an apparent mechanical efficiency of 33 per cent at an ATP concentration of 24×10^{-3} M. This does not, of course, take into account that only part of the bundle could have been active; the true efficiency of the active fibres would clearly be much higher. Even in living muscles the highest mechanical efficiencies, observed in short tetani, are rarely greater than 40 per cent, so that the artificial model system is at least as efficient, per g of active fibres, as the living system and probably more so [24, 71].

Because of the conclusions reached in the last paragraph, the time is evidently now ripe to repeat many of the old observations on this simple system of model fibres, taking the Ca^{++} effect fully into account.

Relaxation and inhibition of ATP-ase activity

In the experiments with isolated fibrils, we showed, that when splitting of ATP was inhibited by chelating out traces of Ca ions, the fibrils no longer contracted in the presence of ATP, but tended to swell: this process we likened to the relaxation of intact muscle. It is very simple to demonstrate the relaxation process with single glycerolated fibres or fibre bundles by chelating Ca^{++} with EDTA or EGTA [12, 152]. In the experiment illustrated in figure 3.4, the fibre bundle was initially loaded at about 1000 g/cm², after washing out the glycerol in buffered KCl solution. In this state, the fibres are in rigor and have a very high modulus of elasticity, so that the load stretches them by only about 2·0 per cent of their length. On replacing the buffered KCl with a solution of 5 mM $MgATP^{-2}$ plus 0·1 mM $CaCl_2$, there is an immediate and rapid contraction amounting to about 50 per cent of the muscle length. At the height of this contraction, the ATP is washed out with KCl solution and shortening immediately ceases. A test of the elasticity of the system at this stage, by applying an extra load of about 500 g/cm², shows that the fibres have once more passed into rigor, and can no longer be easily stretched. This process we envisage as the formation of electrostatic links between the actin and myosin filaments at their points of overlap (see figure 1.3), as soon as ATP is removed from the system. It is obvious that it can occur equally well in the shortened fibres, where the actin–myosin overlap is considerable, as it can at rest length, where the overlap is much less.

After removal of the extra load, we add to the shortened fibre bundle, in its rigor state, a solution of 5 mM $MgATP^{-2}$ plus 2 mM EGTA. As we see, relaxation quickly sets in, and the muscle rather slowly regains the original length it had before the first addition of ATP. We now add sufficient Ca^{++} (2·1 mM), to swamp the chelating effect of the EGTA, and the fibre immediately contracts again. The whole of this cycle, contraction, rigor, relaxation and a new contraction can be repeated several times with the same bundle.

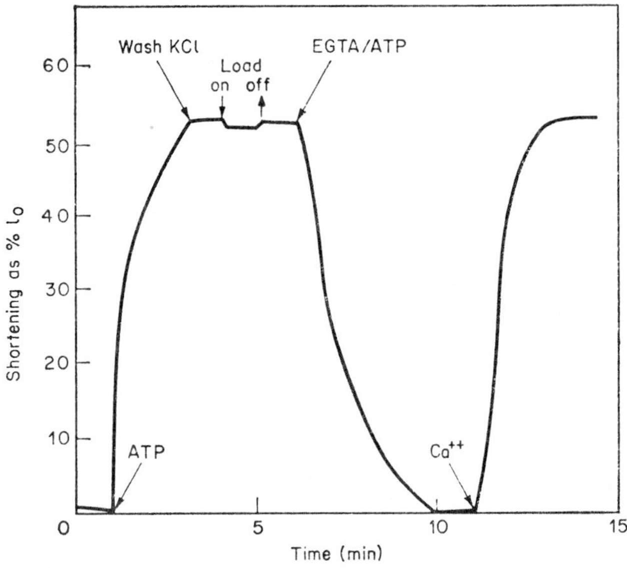

Figure 3.4. Artificial contraction/relaxation cycle in glycerolated fibre bundles. Symbols:

'ATP' = 5×10^{-3} M $MgATP^{-2}$ + 1×10^{-4} M $CaCl_2$
'EGTA/ATP' = 5×10^{-3} $MgATP^{-2}$ + 2×10^{-3} M EGTA
'Ca^{++}' = addition of $CaCl_2$ to final concentration of $2{\cdot}1 \times 10^{-3}$ M.
↓ = extra load of 500 g/cm² added
↑ = ditto removed

(Author's observations.)

Recalling the experiments with fibrils, we now see how the splitting of ATP is related to the contraction–relaxation cycle: in the presence of Ca and Mg ions, ATP is split rapidly by the actomyosin filaments, and the free energy from this process is used for contraction and the development of power, through the mediation of the sliding of the actin and myosin filaments over one another.

Washing out the ATP leads to rigor, but on fresh addition of $MgATP^{-2}$ in the presence of a calcium chelating agent, the ATP-ase activity falls nearly to zero, so that $MgATP^{-2}$ can now act in its role as a relaxant or plasticizer; that is to say, it prevents the formation of electrostatic links between the actin and myosin filaments, which can therefore slide freely over one another on the application of a load (= relaxation).

The same artificial contraction–relaxation cycle can, of course, be demonstrated with single fibres which are developing isometric tension, instead of doing work [160]. The effects of other ATP-ase inhibitors have also been demonstrated in the single fibre system, of which the most interesting in the present context are those which block SH-groups, because this is a quite different type of inhibition from that involving the removal of Ca ions. In such a case we are modifying the enzyme sites on the actin and myosin filaments, instead of merely removing an activator. The complex SH-reagent, salyrgan [159], has frequently been used for this purpose; when it is added, at 1 mm concentration, to a fibre developing tension in the presence of ATP, Mg^{++} and Ca^{++}, the ATP-ase activity and the tension immediately drop to zero and the fibre relaxes. This effect of salyrgan can be reversed by adding 10 mM cysteine, which restores the SH-groups on the enzyme centre once more, and allows ATP to be split and tension to be redeveloped. Since salyrgan has no relaxing effect in the absence of ATP, the above experiments demonstrate once more how $MgATP^{-2}$ always acts as a relaxant, whenever the ATP-ase activity of the actomyosin system is inhibited, whether by this drastic method, or by the more physiological means of removing Ca ions from the system.

ITP as a contracting and relaxing agent

We have seen that actomyosin and fibrils, at physiological ionic strengths, split nucleotides with an NH_2-group in the 6-position of the purine ring, such as ATP, more rapidly than those with an OH-group in that position, such as ITP. It is therefore worthwhile to consider the effect of ITP on glycerolated fibres [12, 152]. A typical experiment is illustrated in figure 3.5, where EGTA is again used as a chelator of Ca ions (cf. figure 3.4). We note first, from the inset to the figure, that the fibre bundle takes some time to react to $MgITP^{-2}$ and to start shortening. The delay here is 8 sec, which is more than 10 times the delay with $MgATP^{-2}$. The initial rate of doing work is also seen to be much lower with ITP than with ATP, on average about $\frac{1}{5}$.

Another essential difference between ITP and ATP is that chelation of Ca ions with EGTA has no relaxing effect in the ITP system,

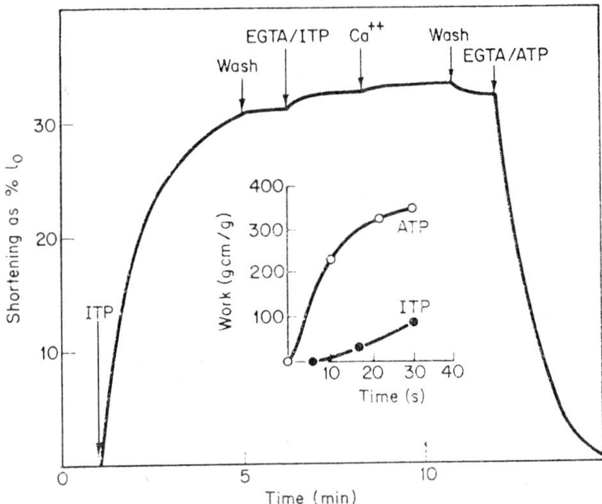

Figure 3.5. Similar experiment to that in figure 3.4, but using ITP as the NTP instead of ATP. [ITP] the same in each case as [ATP] in figure 3.4. Inset to the figure shows the early stages of the work/time curves for 5×10^{-3} M ATP and ITP. (Author's observations.)

as we see from the main figure, where the fibre bundle continues to shorten slowly in the presence of EGTA + $MgITP^{-2}$; moreover, addition of Ca ions in sufficient excess to swamp the chelator has only a slight accelerating effect (cf. [152]). To make sure that these effects were not artefacts, EGTA + $MgATP^{-2}$ was finally added to the bundle and, as we see, produces its normal relaxing effect. It thus appears that ITP under physiological conditions is not an efficient contracting agent, and completely fails as a relaxant. However, in spite of the low work rate in the presence of ITP, there is no doubt that eventually nearly the same total amount of work can be done, and the same total tension developed, as with ATP, but the process takes much longer.

As we said, the initial power developed by muscle bundles in ITP is only about $\frac{1}{5}$ of that in ATP, whereas the rate of splitting of ITP by fibrils is at least half that of ATP [12]. This suggests that the mechanical efficiency in the presence of ITP is $\frac{2}{5}$ that in ATP; a fact which does not seem to have been noticed previously. It is also noteworthy that chelation of Ca ions does indeed reduce the *rate of splitting of ITP* by fibrils but only to about half the maximal value, compared with about $\frac{1}{20}$ with ATP: this no doubt accounts for the lack of relaxing effect with ITP.

The long delay before contraction sets in in the presence of ITP has so far not been explained. A possible explanation arises from the fact that the Michaelis constant for ITP is much higher than that for ATP [61]. Thus, when diffusion is the limiting factor, it will take longer for a sufficient concentration of ITP to build up within the fibre bundle to saturate the enzyme sites.

Another curious feature of the splitting of ITP by fibrils is the extremely low rate when Ca^{++} is used as the sole activating ion; at 35°C for example, it is only about $\frac{1}{15}$ of the rate with $CaATP^{-2}$.

Do fibres contract in the presence of ATP and Ca ions only?

With the fibrillar system, it was obvious that $CaATP^{-2}$ alone was incapable of causing significant synaeresis, despite the fact that it was vigorously split at the enzyme sites (cf. figure 2.5*a* and *b*). Exactly the same occurs with loaded fibre bundles (figure 3.6). Here addition of $CaATP^{-2}$ alone causes no length change whatsoever and the fibres appear to be in state of rigor, since the heavy load they are bearing (1500 g/cm^2) quite fails to extend them. Addition of Mg ions to a concentration of 0·4 mM, however, causes immediate shortening and the performance of work, which cease immediately the Mg ions are washed out and replaced once more

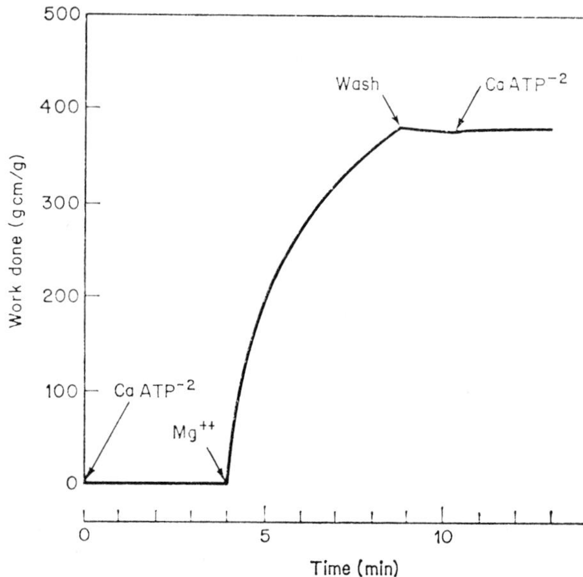

Figure 3.6. Lack of contractile effect when Ca ions are used as the sole activator with glycerolated fibre bundles. Addition of Mg ions immediately induces contraction.

$CaATP^{-2} = 4 \times 10^{-3}$ M
Addition of $MgCl_2$ to 4×10^{-4} M, at second arrow.
Other conditions as in figure 3.2. (Author's observations.)

with $CaATP^{-2}$ only. Even when the load on the fibre bundle is reduced to $\frac{1}{5}$, $CaATP^{-2}$ alone fails to produce a contraction. We conclude therefore that, in spite of ATP being rapidly split in its presence, Ca^{++} is incapable of acting as an activator of contraction under physiological conditions. For this to occur Mg ions are also necessary.

The effect of ageing on the contraction of fibres

There are two interesting effects of age on the contractility of glycerolated fibres, one with very old fibres and the other with very fresh ones. Fibres which have been 'aged' for a period of two years or so at $-10°C$ in 50 per cent aqueous glycerol in the presence of air, contract vigorously when $MgATP^{-2}$ is added to them. As in the case of aged fibrils, however, they are no longer sensitive to EGTA, and completely fail to relax when this Ca chelating agent is added in the presence of ATP. Indeed, they often continue to contract slowly. By analogy with the aged fibrillar system, where the ATP-ase activity can be only partly reduced by chelation of Ca ions (see figures 2.6*a* and 2.7), it seems that here also oxidation of some specific SH-groups of the actomyosin–tropomyosin–troponin complex has occurred during the long storage; these groups are clearly not involved in the contractile process, but only in relaxation; that is, they are responsible for the Ca-sensitivity of the system [20, 115].

An almost exactly opposite effect is noticed with fibres which have been stored in aqueous glycerol for only a day or so; they fail to contract on the addition of $MgATP^{-2}$, even when accidental contamination of Ca ions is as high as 10^{-5} M. Addition of Ca ions to a concentration of about 2×10^{-4} M, however, brings about a rapid and powerful contraction. The experiment can be repeated by washing out the added Ca ions and starting over again, although the relaxing response becomes progressively more feeble on each repetition. This effect was noticed very early on in the history of this aspect of contraction, and some of its implications were already fully realized at the time but later overlooked again [13, 19, 109]. Clearly it indicates that there is some relaxing factor present in fresh fibres, which is lost on ageing, and which is antagonized by addition of Ca ions. Could it perhaps be a Ca-chelator or a Ca pump? Indeed it could, as we shall see in the next chapter.

4: The Sarcoplasmic Reticulum and Relaxation *in vivo*

Introduction

The last two chapters have demonstrated the extreme sensitivity of the actin and myosin filaments of natural fibrils to Ca ions, under physiological conditions of high [ATP] and [Mg^{++}]; probably necessary for this sensitivity is the presence of tropomyosin and troponin within the double-stranded helices of actin. We concluded that this ability of very low concentrations of Ca ions to push the system over from a state of rest to one of full contractile activity, that is from a very low 'resting' ATP-ase activity to a very high one, could form the basis of the coupling of excitation to contraction in the living system. All that would be necessary would be a calcium pump to keep the free Ca concentration low during rest, and a system for inhibiting this pump and allowing Ca ions to diffuse from it to the active enzyme sites, when the electrical excitation wave, or action potential, passed down the muscle membrane, via the nerve and motor end-plate.

The history of the discovery of the natural calcium pump as part of the complex system of longitudinal and transverse tubules and vesicles which enwrap each fibril and which are known collectively as the sarcoplasmic reticulum, is an extraordinary one, full of unexpected twists and turns. Indeed by discussing the effect of Ca ions first and the pump afterwards, we have reversed this history, because the first natural factor to be discovered, the so-called Marsh factor [109], although composed of fragments of this very reticulum, was thought to effect relaxation by quite different mechanisms [118]. This was perhaps inevitable, since the factor was prepared by extracting muscle homogenates at low ionic strength, and therefore contained not only reticular fragments, but also all the enzymes of anaerobic glycolysis, with some of which it was confused. It was not until high speed centrifuges became easily available that the apparently clear factor suspensions were shown to contain not only broken up mitochondria, but also fragments of the reticular vesicles [21, 128] (plate IV.2). It was the latter which possessed relaxing ability.

However, even when the relaxing factor had been shown to be particulate in nature, its Ca pumping ability was not recognized for some years, although it was already known that low concentrations of Ca ions reversed its effect [39, 40, 62, 109, 156]. This discovery also had to await technical developments, particularly refined methods of estimating Ca, and of using radio-calcium (Ca^{45}). Moreover, the sarcoplasmic reticulum itself, although discovered by light microscopy in the early years of the century [151] and described in great detail, had been completely overlooked for more than 50 years by muscle histologists and physiologists alike, and was dramatically rediscovered by electron microscopy in the midst of all the investigations on contraction and ATP-ase activity we have discussed [126]. Its impact has been decisive, and we shall therefore describe its structure before discussing its action as a Ca pump.

The sarcoplasmic reticulum

A simple way of envisaging the structure of the reticulum in longitudinal section is shown diagrammatically in figure 4.1 [131, 132]. On the left is shown the plasmalemma or semi-permeable membrane of a muscle fibre: this is characteristically triple-layered, and about 100 Å thick [120]. Like the plasmalemma of all other cells, this membrane maintains the delicate osmotic equilibrium of the muscle-fibre, and also serves the very important function of conveying the action potential along the fibre, from its nervous origin at the motor end-plate [76, 95]. But, we may enquire, how is this action potential propagated inwards to the innermost fibrils within the fibre, which are often at least 25 μ (250,000 Å) away from the

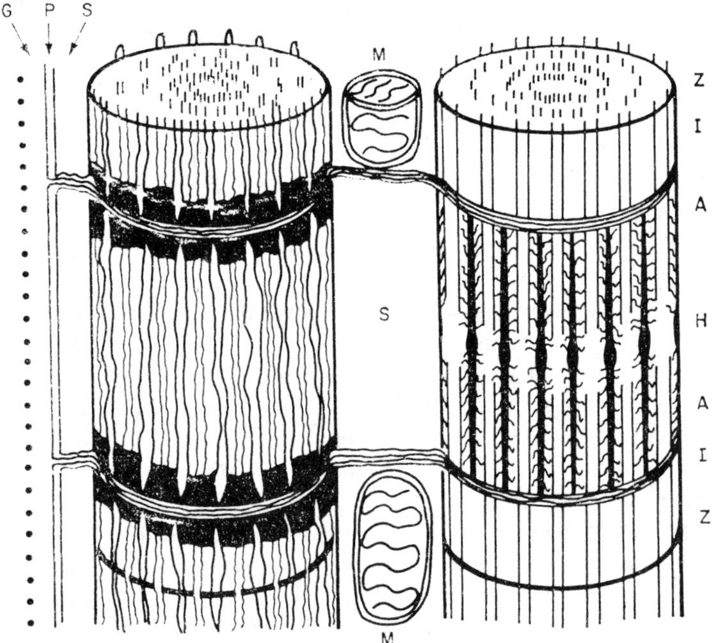

Figure 4.1. Diagram of two fibrils of a rabbit muscle to illustrate ramifications of sarcoplasmic reticulum (SR). Only one complete sarcomere is shown in each case, the left-hand one with SR intact, and the right-hand one partly stripped to show underlying myofilaments. The transverse elements of the SR can be seen arising from the plasmalemma (P). Note the dark swellings at the ends of the longitudinal vesicles; these are the terminal cisternae (TC) of the triads, which are situated in this case at the A–I boundary. S = sarcoplasm; M = mitochondria; G = ground substance.

membrane? The answer lies in the so-called transverse tubules of the reticular system, which as we see originate as invaginations of the plasmalemma itself [49, 84], and continue inwards as tubules which enwrap each fibril at the level of the junction of the A- and I-bands in mammalian and some fish muscles, and at the level of the Z-lines in many cold-blooded species [126, 131] (see also plate IV.1 *a–d*).

Although not described in detail in most studies of the reticulum, it is apparent that the transverse network of tubules must connect across from fibril to fibril, at the same level in each sarcomere. The

Plate IV.1a The sarcoplasmic reticulum in frog sartorious muscle, after treatment with ferritin. Note the small dark particles of ferritin in portions of the transverse tubules, running above the Z-discs, and the thickened vesicles of the triads on either side of these dark tubules. The triads connect with the irregular longitudinal vesicles, running parallel to the contractile filaments. These contain glycogen granules. (Photo by courtesy of Dr H. E. Huxley.)

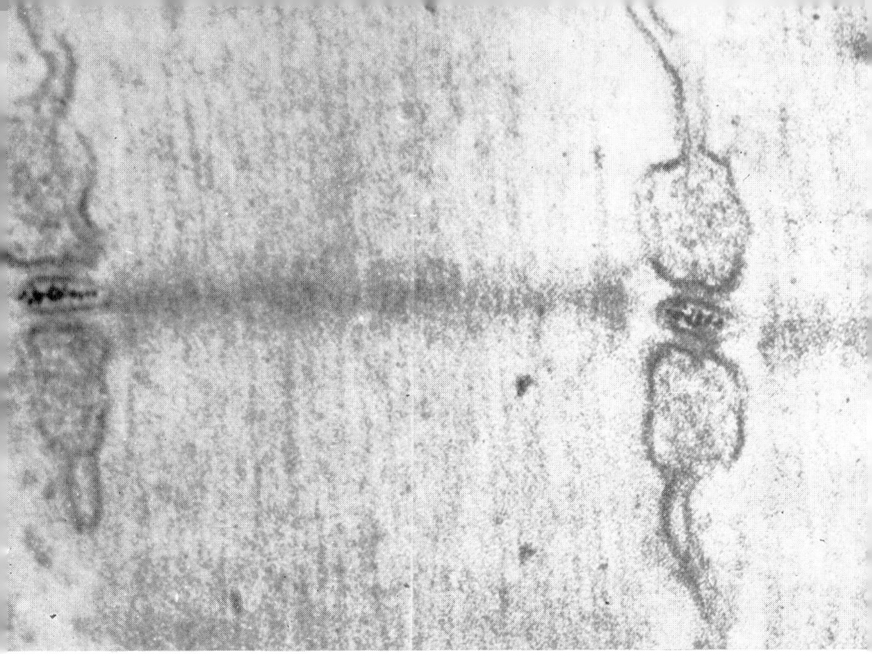

Plate IV.1b Higher magnification views of the triad region of the sarcoplasmic reticulum. Note how the triad structure sits exactly on top of the Z-disc in this muscle (frog sartorius). Also note that ferritin has here penetrated only the central, transverse element of the triad, a portion of which has been cut in glancing section. The longitudinal elements are free of ferritin.

Plate IV.1c A longer run of transverse tubule, with longitudinal vesicles on either side of it. Note how the transverse tubule runs above the Z-disc.
(Photos by courtesy of Dr H. E. Huxley.)

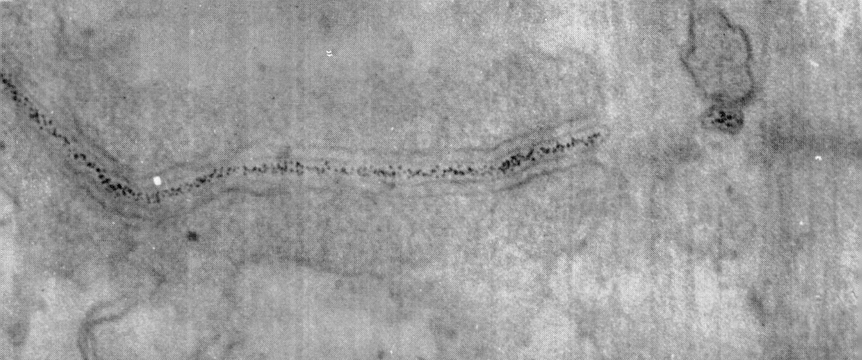

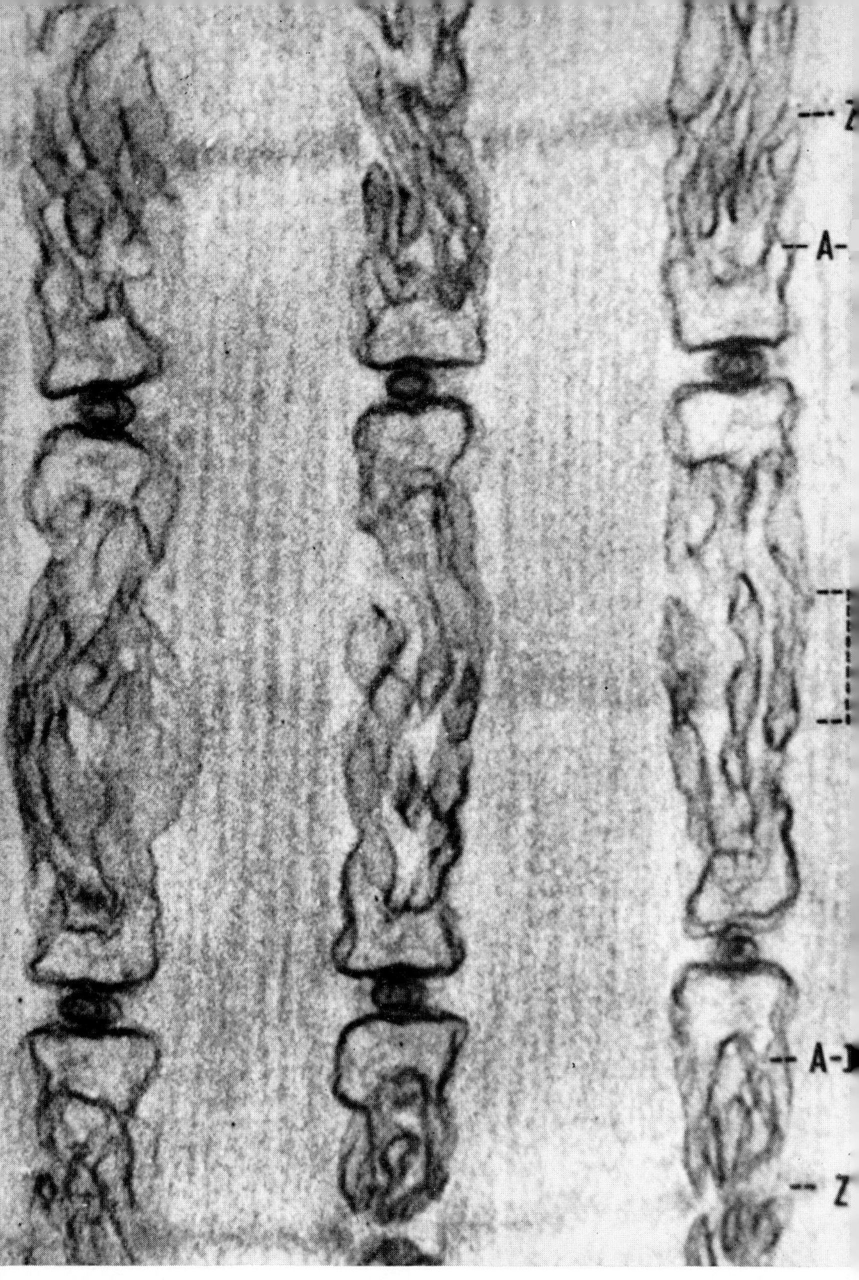

Plate IV.1d Glancing section of the sarcoplasmic reticulum of a fish muscle (toadfish swim bladder). The transverse tubules appear as a dark circle, lying near the A-I junction and *not* on the Z-disc as in the frog. The triads can be clearly seen, with longitudinal vesicles running away in either direction, parallel to the underlying myofilaments. Note that this plate is very similar to the diagram of a mammalian muscle (fig. IV.1).

(Photo by courtesy of Drs D. W. Fawcett and J. P. Revel.)

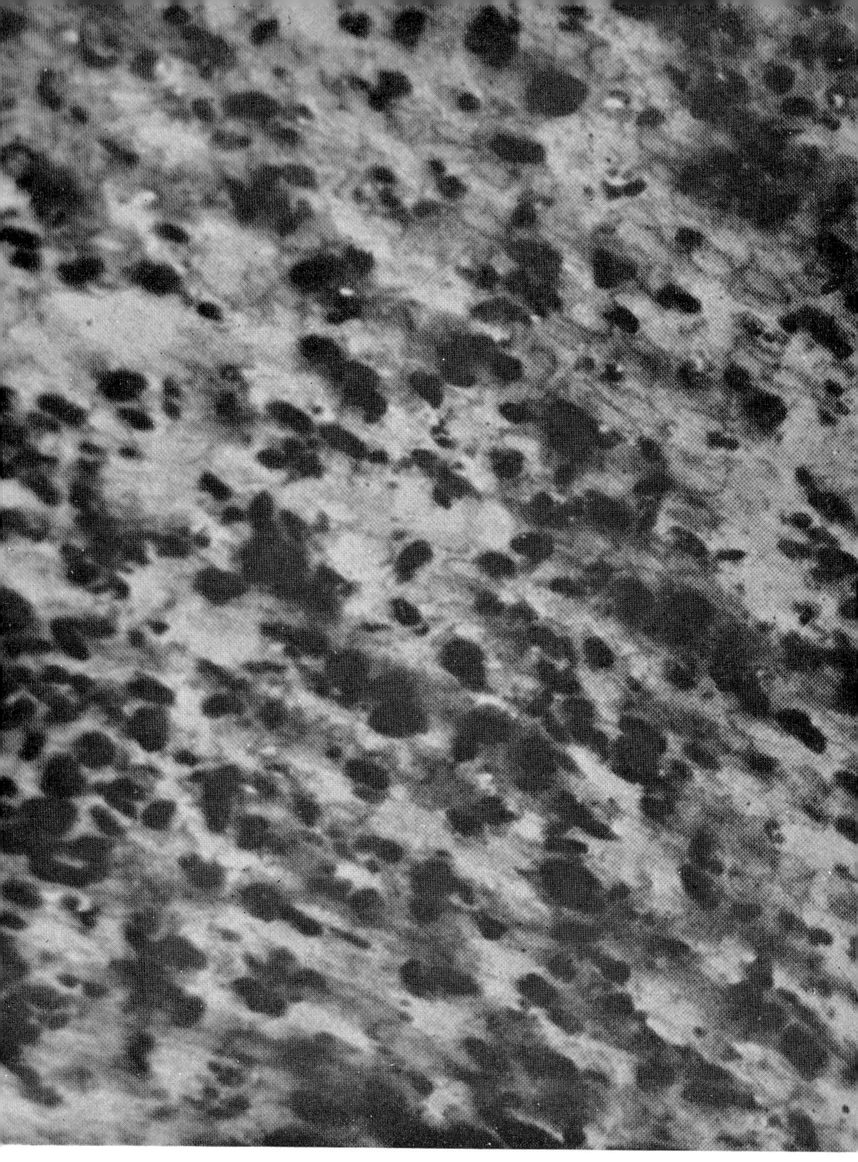

Plate IV.2 Preparation of sarcoplasmic reticular fragments from rabbit muscle, after homogenization. The fragments have been treated with Ca^{++}, which has been taken up by some of them (dark particles).

(Photo by courtesy of Dr W. Hasselbach.)

complexity of the net can be partly envisaged from figure 4.1, where only two of the outermost fibrils of a fibre are shown, one with its sarcoplasmic reticulum intact, and the other stripped to show these ramifications in relation to the underlying actin and myosin filaments. In reality, most muscle fibres contain at least 1000 fibrils, each of which must be interconnected in this way; and not only that, because in the flight muscles of birds and insects, for example, mitochrondria, associated with the oxidative metabolism of the fibre, are packed in rows between the fibrils, so that the transverse tubules of the reticulum must also pass this gap on their way inwards. Add to this that in a 1-cm length of fibril there are some 4000 sarcomeres, transversely connected to their neighbours, and we see that the transverse tubules of the reticulum are sufficiently complex and numerous to explain the oft observed fact that the cross-striations of all the fibrils within a fibre stay beautifully in register, even during a vigorous contraction.

Before leaving this aspect of the subject, it should be mentioned that the transverse connections with the plasmalemma were first shown up by soaking intact muscles or muscle fibres in suspensions of very fine ferritin particles, which can enter the transverse tubules through the pores in the plasmalemma, as shown in plate IV.1 *a–c* [49, 84]. Note that there is no penetration of ferritin into the longitudinal vesicles (LV), showing that these are not directly connected to the transverse system; the particles in the LV are glycogen.

We can think of the transverse system as consisting of very thin tubules, about 400 Å in diameter, but it is more difficult to envisage the structure of the LV (longitudinal vesicles), although plates IV.1*a* and *d* give some idea of their complexity. In fact the LV totally enwrap the fibril, running in mammals and some fish from the triads at the A–I junctions, in both directions along the fibril, and in many cold-blooded species such as the frog, from Z-disc to Z-disc (cf. plates IV.1 *a–c* and plate IV.1*d*) [49, 84, 131]. At their termination at the A–I junctions in the former case, these vesicles form swellings which are in contact with the transverse tubules, and this whole system is known as a triad, because it consists of two of these swellings, or terminal cisternae (TC) on either side of a transverse tubule (see figure 4.1 and also the glancing section in plate IV.1*d*.)

It is across the minute gap between the transverse tubules and the TC of the triads that the action potential must be conveyed. We may now enquire what chemical function each part of this system performs.

Ionic pumps

Without involving ourselves at this stage in an exact description of the action potential, we can say that a nervous impulse passes down a motor nerve to the motor end-plate of a muscle fibre in the form of a depolarization wave or action potential; there it releases acetyl choline from the end-plate, and this in turn starts off another action potential along the muscle membrane [76, 95]. During this process a change in semi-permeability occurs locally so that a very small number of Na ions can leak into the muscle or nerve fibre, and an equal number of K ions can leak out. Even though the exchange of ions per stimulus is small, it is obvious that there must be some system available for eventually restoring the *status quo* at the end of a series of stimuli, and pumping the Na and K ions back to their original quarters. Na/K pumps of this sort have been extensivley investigated in recent years, and turn out to be run, in a complex way, on the free energy from ATP breakdown [120]. Because the number of ions to be transferred back across the membrane during recovery is so small, the amount of ATP used in this process is much too slight to be detectable amongst the massive energy fluxes which occur during a contraction [76].

Although not proven, it follows logically from the fact that the transverse tubules of the reticulum are really invaginations of the plasmalemma that they too convey the action potential within the fibre in the form of an Na depolarization wave, and that this wave sets off a similar form of impulse across the two gaps in the triad system, and thus in either direction along each fibril. However, the ions involved in longitudinal transmission are no longer Na^+ and K^+, but Ca^{++} and possibly Mg^{++} [39, 40, 62, 63]. This we know from the original studies of the Marsh factor which was shown to consist of small fragments of reticulum, by centrifuging these down at high speed from the apparently clear solutions which contained them, and examining the precipitate in the electron microscope [39, 40, 62]. This precipitate consists of short lengths of fine tubules, and also of structures which can be recognized as parts of the LV and triad (TC) system (see plates IV.1*d* and IV.2). Note that it is only the latter which are capable of taking up Ca^{++} from the medium, as shown by their much denser appearance in plate IV.2. The longitudinal vesicles (LV) or more probably the TC of the triads can indeed operate quite efficiently as a calcium pump, but show no tendency to pump Na or K ions [63]. As we shall see, this Ca pump is also run on the free energy from the hydrolysis of ATP.

ATP and the Ca pump

The usual Marsh factor preparations, after partial purification by differential centrifugation, split ATP rather slowly in the absence of Ca ions, but when even quite small quantities of Ca^{++} are added to the suspension, the ATP-ase activity rises sharply [39, 40, 62, 156]. This so-called 'extra' ATP-ase activity can be shown to result in the accumulation of Ca^{++} in the reticular fragments, by using radioactive Ca (Ca^{45}), and 'counting' the centrifuged precipitate. For Ca accumulation to approach the levels expectable *in vivo*, however, some substance must be added which will precipitate as its Ca salt within the more or less leaky vesicular membrane [63, 156, 157]. Oxalate is usually employed for this purpose, but may be replaced by phosphate; it can be shown that both these substances accumulate within the vesicles by free diffusion, and precipitate there as their Ca salts, as the internal Ca^{++} concentration rises [40, 62, 63, 156, 157]. For this reason, it has been postulated that inorganic phosphate may act in this fashion *in vivo*, although there is no direct experimental evidence for it [31]. With oxalate present, the Ca pump is capable of concentrating Ca^{++} within the vesicles to 2000 to 3000 times the external Ca concentration. Even without oxalate, Ca can be concentrated to at least 500 times the external level [40, 63, 157].

The early observations on the pump suggested that between 1 and 2 Ca ions were accumulated within the vesicles per ATP molecule split [62, 63], but more recent studies show that the correct ratio is 2 Ca ions per ATP, independent of the external Ca concentration [157]. Figure 4.2 illustrates the relation between the rate of Ca uptake and of ATP splitting, at varying external Ca concentrations, from which we see that a line drawn with a slope of 2 fits the points

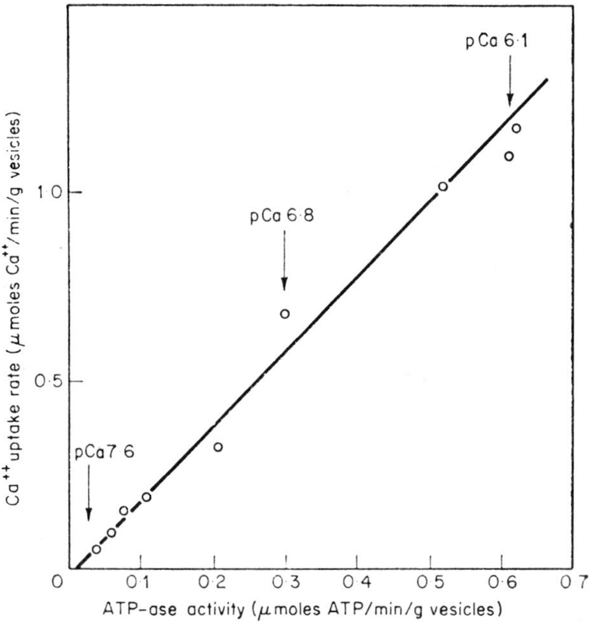

Figure 4.2. **Plot of the ATP-ase activity of suspensions of sarcoplasmic reticular vesicles against their ability to pump and store Ca ions.**

Conditions: Ionic strength = 0·1; temp. = 20°C; CaEGTA buffers. [ATP] = 10×10^{-6} M. PC (2×10^{-6} M) and PC-kinase added to keep [ATP] constant. The line is drawn with a slope of 2. After [157].

reasonably well. The external 'free' Ca concentration was maintained constant in most of the experiments to be described by using Ca-EGTA buffers [157]. As we have noted these Ca buffers are in every way similar to the more usual pH buffers, except that they keep the free Ca^{++} concentration constant, instead of that of H^{+}. The vesicles themselves are impermeable to EGTA, so that the salt, Ca-EGTA, does not enter to disturb the internal Ca equilibrium.

Because ATP can also chelate Ca ions, it might be thought that the pump transported these ions through the vesicular membrane in the form of the Ca–ATP chelate, but this is definitely not so, since the initial rate of Ca uptake is independent of the concentration of chelate over a wide range [157]. On the other hand, the uptake rate depends closely on the external concentration of free Ca ions, as shown in figure 4.3, where the latter parameter is plotted on the abscissa as pCa (= negative logarithm of the Ca concentration).

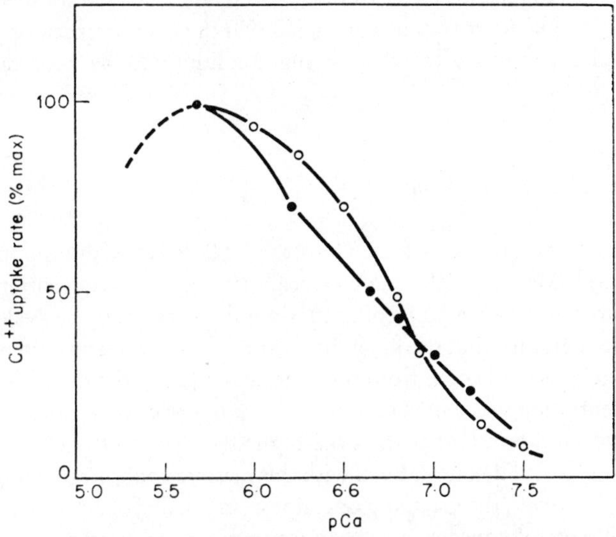

Figure 4.3. Effect of external [Ca^{++}] on the Ca^{++} uptake rate by sarcoplasmic vesicles. Conditions similar to figure 4.2. (After [157]).

● uptake rate by empty vesicles.

○ uptake rate by empty vesicles in presence of oxalate.

Since the uptake rate is dependent in its turn on the ATP-ase activity (cf. figure 4.2), it follows that the latter is solely determined by the external Ca concentration, in the initial stages when the vesicles are empty. Figure 4.3 shows that there is some difference in the slope between the two curves relating uptake to pCa, depending on whether the rate is measured in the presence of oxalate, and thus with a very low internal Ca concentration, or by using 'empty' vesicles in the absence of oxalate. It is doubtful whether this difference is significant when all the data are considered.

Although the uptake rate and rate of ATP splitting are related to the external Ca concentration in a sharp and clear cut manner, this is not so when the effect of ATP concentration is considered [157]; however, it is at least obvious from the somewhat anomalous results that very high rates of uptake can still be obtained with ATP concentrations as low as 10^{-6} M. Such low ATP concentrations are rarely encountered in intact muscle, even after the establishment of rigor, and hence it is to be concluded that the Ca pump can operate at full capacity under most of the conditions to be expected *in vivo* [63].

Other features of the pump

Besides a requirement for Ca ions, the ATP-activated pump appears to need Mg ions, although critical data for the concentrations required at varying ATP concentrations do not seem to be available in the literature (but see [63]). In most of the experiments discussed above, which all come from one laboratory [157], the total Mg ion concentration was 1 mM or above, that is to say two or more orders of concentration higher than the highest Ca concentrations added to the external medium. Even with this massive concentration difference, however, the Ca pump does not seem capable of actively transporting, or of concentrating, Mg ions within the vesicles. The role of Mg ions is thus obscure, although it may be that they merely serve to keep the free [ATP^{-4}] within reasonable limits, by chelating the excess as $MgATP^{-2}$. *In vivo*, of course, the total concentration of Mg ions is at least 8 mM in most mammalian muscle. Mn ions can replace Mg in this respect. Strontium in contrast to Mg and Mn *can* be concentrated by vesicles, almost as efficiently as Ca [63, 157].

The importance of Mg to the pumping can be demonstrated in another important respect, the phosphate exchange which occurs during the process, although here too it is by no means certain that Mg is playing an active role, it can again be replaced by Mn. Thus, both these ions probably act by reducing the free ATP^{-4} level, which might otherwise become over-optimal.

The phosphate exchange which occurs during pumping does so by some such mechanism as the following:

$$\left.\begin{array}{ll} (a) & \mathrm{NTP} + \mathrm{E} \underset{k}{\overset{k_1}{\rightleftarrows}} \mathrm{E} \sim \mathrm{P} + \mathrm{NDP} \\ (b) & \mathrm{E} \sim \mathrm{P} \underset{k_3}{\longrightarrow} \mathrm{E} + \mathrm{P} \end{array}\right\} \quad 4.1$$

where NTP may be ATP itself, or GTP, ITP or CTP, and E is the enzyme site [63, 105]. In the scheme, we suppose that k_2 and k_3 are not too much greater than k_1, so that the complex, E $\sim$ P, can exist for a finite time and hence react with any NDP present, to give the corresponding NTP. Thus, if we start with, say, ITP labelled with radioactive-phosphorus in its terminal, γ-phosphate group, and then add unlabelled ADP, the following exchange reaction occurs, from which a small proportion of labelled ATP can be isolated by paper chromatography:

$$\left.\begin{array}{ll} (a) & \mathrm{ITP}^* + \mathrm{E} \rightleftarrows \mathrm{E} \sim \mathrm{P}^* + \mathrm{IDP} \\ (b) & \mathrm{ADP} + \mathrm{E} \sim \mathrm{P}^* \rightleftarrows \mathrm{E} + \mathrm{ATP}^* \end{array}\right\} \begin{array}{l} \text{and also loss of label through:} \\ \mathrm{E} \sim \mathrm{P}^* \overset{k_3}{\longrightarrow} \mathrm{E} + \mathrm{P}^* \end{array} \quad 4.2$$

Clearly the degree of exchange depends on the nature of the NDP used, which will determine the magnitude of the rate constant, k_2, in the first equation. Furthermore, we do not even need to use different NTP's, since ADP labelled in its terminal, β-phosphate group can be added instead of unlabelled ADP and by exchange with unlabelled ATP, we shall then end up with some ATP labelled in its β-phosphate group.

Apart from the interest which attaches to the relative rate constants of the various NTP's, given in [105], the exchange reaction has the following important features: (i) it occurs only in the presence of Ca ions, and therefore it is involved in the Ca-pumping activity; (ii) labelled inorganic phosphate does not exchange, and therefore, reaction 4.1*b* is virtually irreversible; (iii) it reveals part of the mechanism of pumping, since it could not occur at all, except through a phosphorylated intermediate which is either the enzyme site itself or a carrier.

Structural features

The longitudinal vesicles contain considerable quantities of lipid material, mostly in the form of phosphatidylcholine and choline plasmalogen, with some phosphoinositide and phosphatidylethanolamine [110]. This lipid is essential for the pumping action, which ceases entirely when it is removed by detergents or phospholipases, although the ATP-ase can still operate [110]. Pumping activity can be restored by adding lecithin and other lipids. These experiments therefore suggest that the vesicle membrane, like other plasma-membranes, consists of interleaved lipid and protein constituents, with the charged groups of the lipid covered with a protein monolayer which sticks out into the aqueous medium, and the fatty acid side-chains buried within the structure. It is probably rather premature to take this argument further, and attempt to draw a detailed structure, but excellent generalized drawings of other types of membrane are given in [120].

Another feature of the LV, or more probably TC, membrane, which is characteristic of all enzymes which utilize ATP, is the presence of several different species of SH-groups, some of which are clearly involved in the ATP-ase and pumping activity [64]. The total number of SH-groups is about 10 per 10^5 g of protein, 4 being directly involved in the pumping activity, which is lost when they are blocked by N-ethyl maleimide (NEM) or salyrgan (a complex salicyl-mercuri-compound). Increasing blockage reduces the ATP-ase activity, the Ca-uptake rate and the concentrating ability proportionately, but storage ability does not appear to be affected until more than 90 per cent of the groups are blocked. In other words, the vesicles will fill satisfactorily at low rates of ATP-ase acitivity or of Ca uptake, whether these low rates are brought about by a low external Ca concentration or by blockage of SH-groups. Addition of ATP, before an SH-reagent is added, affords protection against blockage, just as it does in the case of the fibrillar ATP-ase (see chapter 2). As expected from the exchange reaction discussed above, ADP will also protect against SH-reagents.

By the use of an electron dense stain, Hg-phenylazoferritin, the

SH groups involved in Ca transport can be located on the outer surface of the LV or TC membrane, as shown in plate IV.4.3 [65].

The Ca pump in vivo

The results we have so far discussed were all obtained on fragments of the sarcoplasmic reticulum of rabbit or frog muscle, where the active fragments almost certainly came from the TC of the triad system at the A-I junctions or Z-discs. We may now ask what evidence there is that a Ca pump actually operates in living muscle. Until very recently, the proof of this rested entirely on the injection of Ca^{++} ions into living fibres, where the free $[Ca^{++}]$ was controlled by means of Ca/EGTA buffers [129]. In the giant muscle fibres of the crab (*Maia squinado*), for instance, the injection of relatively high concentrations of a Ca buffer, calculated to give a final concentration of free Ca^{++} ions of the order of 5 to 10×10^{-7} M, stimulated full contraction within about 5 sec of injection. On the other hand, the fibre failed to contract when the concentration was reduced to 10^{-7} M [129]. Reference to figure 2.6*a* shows that this agrees well with the more detailed results for the effect of varying $[Ca^{++}]$ on the myofibrillar ATP-ase activity.

A most interesting feature of the injection experiments was that the injected fibres always began to relax again about 20 sec after a full contraction, suggesting that the free Ca^{++} level had been reduced by a Ca pump from an initial level of about 10^{-6} M to below 10^{-7} M. At first sight, this is not a very impressive amount of pumping, but we must remember that to reduce the free Ca^{++} level in this way, it is necessary to remove a much larger amount of Ca^{++} from the CaEGTA buffer which was initially present at a concentration of about 3×10^{-3} M, and which would progressively dissociate as the free Ca level was reduced by the pump. In fact, about $1{\cdot}7 - 10^{-3}$ moles of this would have to be removed per kg of muscle to reduce the free Ca^{++} level below 10^{-7} M. If the crab fibres contained about the same number of triads as rabbit, this result could be easily explained on the basis of the *in-vitro* studies of the pump we have outlined above. More direct evidence for the release of Ca^{++} on stimulation of a living muscle, and for its reabsorption during relaxation, has recently been obtained by an elegant photometric technique, after ingestion of a Ca-indicator directly into the muscles (figures 6.2 and 7.2 and reference 90).

Distribution of Ca within intact muscle

The problem of the distribution of Ca within living muscles is important, because it is obvious from all we have said that a resting muscle cannot tolerate a concentration of free Ca ions greater than 10^{-7} M, yet the total Ca content of most muscles is of the order of 1 to 2×10^{-3} M [63]. How much of this Ca is accumulated within the sarcoplasmic reticulum, and how much is bound to other structures? These other structures are chiefly the actin and myosin filaments and the mitochondria.

Actin contains one very firmly bound Ca ion per molecule (MW = 47,000), whereas myosin appears to contain two, one of which is easily exchangeable with Ca^{45} and the other not [7, 153]. Troponin can also bind Ca^{++} to the extent of about 4 ions per molecule of 80,000 MW, and this Ca is also exchangeable [20, 35*a*]. The myofibrillar proteins consist of about 27 per cent actin, about 54 per cent myosin and about 4 per cent troponin, and their total content per g of wet muscle is 0·12 g [57]. From this, the following molar values are obtained for the Ca binding by these components in one kg of wet muscle: 7×10^{-4}, $2{\cdot}6 \times 10^{-4}$ and $2{\cdot}5 \times 10^{-4}$, respectively. This would account for a total Ca content of about $1{\cdot}2 \times 10^{-3}$ moles per kg, which is close to the value for the total Ca content of rabbit muscle; hence, binding by mitochondria and other particulate structures must be zero or very small indeed. Of the Ca bound to the major structures, only about $0{\cdot}40 \times 10^{-3}$ moles per kg are considered to be easily exchangeable, and thus, easily removable by the action of the Ca pump. The maximum storage capacity of the pump is, however, of the order of 30×10^{-3} moles per kg of muscle by direct measurement. This gives a large safety factor, at least in the special case of rabbit myofibrils, even though the storage capacity was measured in the presence of oxalate, which would tend to increase it [63].

The Ca pump in non-skeletal muscles and slow muscles

The distribution and detailed structure of the sarcoplasmic reticulum varies widely from muscle to muscle, but it seems to be a general rule that the faster acting the muscle the more extensive its reticulum [63, 126, 131]. We might, therefore, expect that slow muscles such as those of the heart and the even slower ones of the gut and arteries would contain relatively little reticulum. This is indeed so, and it is only comparatively recently that it has become possible to isolate an active preparation of the Ca pump from heart muscle [158], and many smooth muscles either contain none or else very little [63]. Because muscles of the latter type certainly do not lack Ca, it has been suggested that the free Ca^{++} concentration within them is controlled not by a reticulum, but by the plasmalemma itself. In other words, Ca^{++} is pumped inwards and outwards of the whole muscle cell during activity [63]. Such a mechanism could, of course, operate only in exceedingly slow muscles, because it would clearly upset the delicate Na^{+}/K^{+} balance in the plasmalemma, on which the depolarization and repolarization of fast, skeletal muscles depend; it would also present a very long diffusion pathway.

Possible mechanisms of Ca transport

At present the exact mechanism of transport of ions through any biological membrane is a matter of speculation, and although more or less complex carrier mechanisms have been postulated, none of the hypothetical carriers has yet been isolated. In common with the mitochondrial membrane, for instance, it has been suggested that Ca transport across the TC membrane involves two carriers [157]; while this mechanism certainly explains some of the facts, it introduces confusion into other observations. In view of this, it is premature to go too deeply into the subject. Nevertheless one cannot help feeling a certain reluctance to accept the presence of unidentifiable entities, when a simpler mechanism might equally well explain the phenomenon. For example, it is certain that when ATP reacts with the ATP-ase sites of myosin during a living contraction, it induces shape changes in the heads of the myosin molecules which protrude from the thick filaments, and that it is these changes which result in the active sliding of the actin and myosin filaments over one another during contraction, with the performance of mechanical work or development of tension [82, 83]. During this process there must be a considerable transfer of Ca^{++} and Mg^{++} from one site to another. It is therefore not too fanciful to suppose that the active ATP-ase centres on the surface or within the membrane of the reticular vesicles can also go through similar ATP-induced changes of shape during their ATP-ase and Ca-pumping activity, and that it is this local shape change which results in the accumulation of Ca ions within the vesicles. For such a process no carrier would be needed, but only a certain degree of flexibility of the protein chains in the region of the active ATP-ase and Ca-binding sites, which could turn themselves more or less inside out, or outside in, thus carrying the attached Ca^{++} with them and releasing it on the inside of the membrane, each time an ATP molecule was split. This suggestion might also explain the observed exchange of phosphate between ATP and ADP. There is no space here to go further into this problem, but a similar mechanism has recently been proposed for mitochondria [171].

Another most important point about the mechanism of pumping is its exact location: is it situated all along the LV or restricted to the terminal cisternae (TC) in the triad region? This question has not yet been satisfactorily resolved, but on general grounds the terminal cisternae seem the most likely location of the pump, and therefore also for release of Ca^{++}, when the AP arrives in the triad region [66]. Otherwise it is difficult to see the reason for the adaptation, peculiar to fast mammalian and fish muscles, of doubling the number of triads and cisternae within a sarcomere by placing them near the A-I junctions, instead of at the Z-lines (figure 4.1 and plates IV.1*c* and *d*). In what follows we shall assume that this is indeed the case: the pump is situated at the terminal cisternae (TC).

Part Two: Intact Muscle

5: Transmission of the Motor Impulse

Although in most of the studies of contraction in living muscle which we shall describe here, the muscle was stimulated directly and not through its nerve, it is important to know how transmission of the motor impulse occurs in the living animal because this illustrates several important general features of conducting membranes. The simplest way of envisaging this process is to consider a single motor unit, as illustrated diagrammatically in figure 5.1 [28, 29].

We see from the figure that each motor nerve fibre within a nerve trunk arises from a single nerve cell, or motor neurone, in the ventral horn of the spinal cord, and runs in the trunk until it reaches the muscle it is to activate. There it breaks up into fine branches, each of which terminates in a group of nerve terminals, called a motor end-plate, in contact with a single muscle fibre. A single nerve fibre often gives rise to fifty or more of these branches, each terminating in an end-plate. The whole group is referred to as a motor unit.

The motor units themselves are frequently grouped towards the centre of the muscles they activate, as shown in plate V, for the small biceps brachii of the mouse. This is clearly the most advantageous position, since the impulse must be conveyed as rapidly as possible along the muscle fibre to activate it fully. It is probable that in most muscles the motor units correspond to the primary muscle bundles, illustrated in plate A, each containing between twenty and sixty muscle fibres, delineated by the so-called perimysial connective tissue.

The central core, or axoplasm, of the nerve fibre and of its branches, and the cell membrane which surrounds it are together known as the axon. The membrane of the axon is the conducting part of the fibre, and can be considered as a very long extension of the membrane of the motor neurone of origin: in other words, a nerve fibre with its neurone and terminal branches is really a single cell with a continuous membrane, the nuclei lying in the neurone itself. It is important not to become confused between this mem-

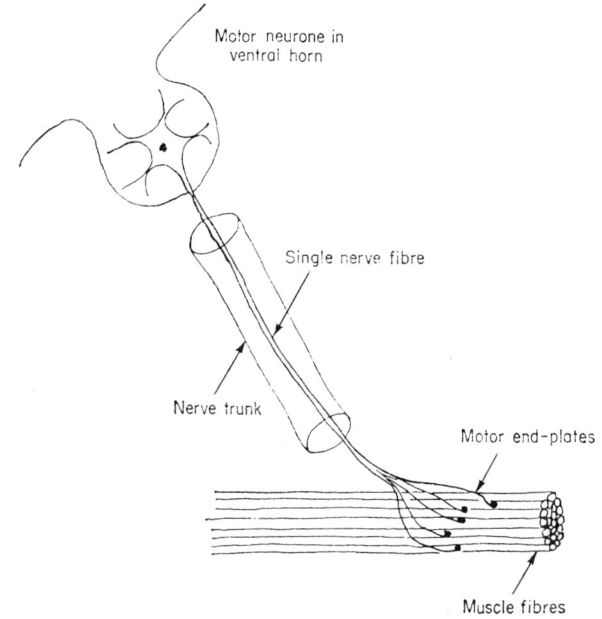

Figure 5.1. **Diagram to illustrate the origin of a motor nerve fibre from its neurone in the ventral horn of the spinal cord, and how it runs in the main nerve trunk. It finally gives off branches to a group of muscle fibres (motor-unit), each branch terminating in a motor end-plate on a single muscle fibre.**

brane which is characteristically triple-layered and only about 100 Å thick, and the much thicker, outer protecting layer of Schwann cells.

In the primitive, *non-myelinated*, fibres of the autonomic nervous system, Schwann cells make only a simple protective coat, usually open on one side, as shown in the cross-section of an axon in figure 5.2*a*. On the other hand, in the *myelinated* motor nerve fibres of the 'voluntary' system, they spin complex, concentric rings of myelin around the fibre, and these totally enclose it in the broad regions between the nodes of Ranvier, shown in figure 5.2*b*. In the internodes such fibres are highly insulated against the external medium,

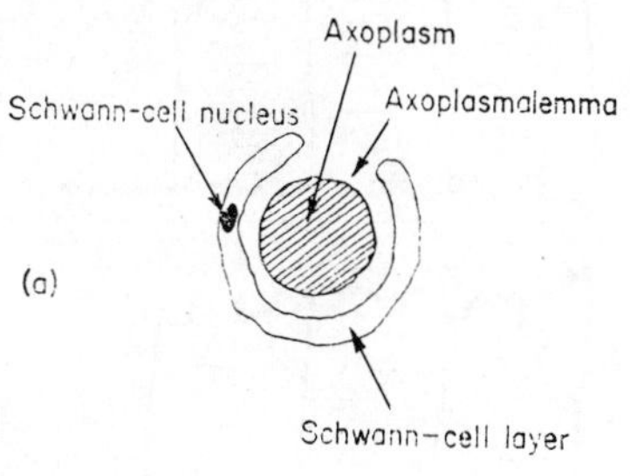

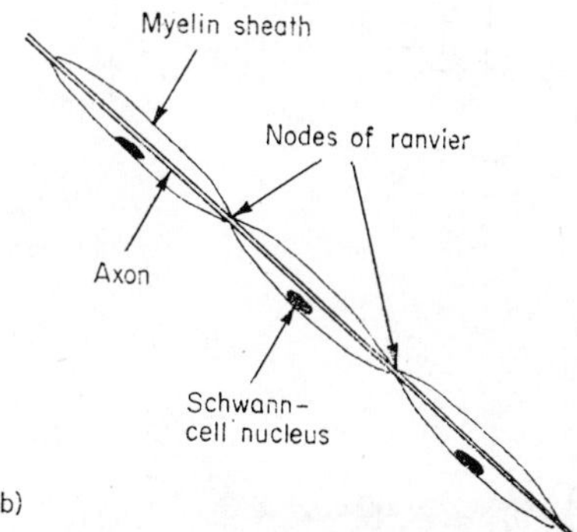

Figure 5.2. (*a*) Diagram to illustrate the manner in which Schwann cells spin a coat round so-called non-myelinated nerve fibres (cross-section).

(*b*) Diagram of the complex myelin sheaths spun by Schwann cells round myelinated fibres (long section). Note the nuclei of the individual Schwann cells, in the highly insulated internode region, and the bare patch of nerve fibre at the node of Ranvier (after [76]).

whereas in the nodal region the true nerve fibre membrane itself is bare. The internode region is 1 to 2 mm long, and contains a single Schwann cell, whereas the node is less than 0·1 mm. The diameter of myelinated fibres usually lies between 1·0 and 20 μ (0·001 and 0·02 mm). In some animals, such as the squid, Loligo, giant nerve fibres occur as large as 1 mm in diameter, but they are non-

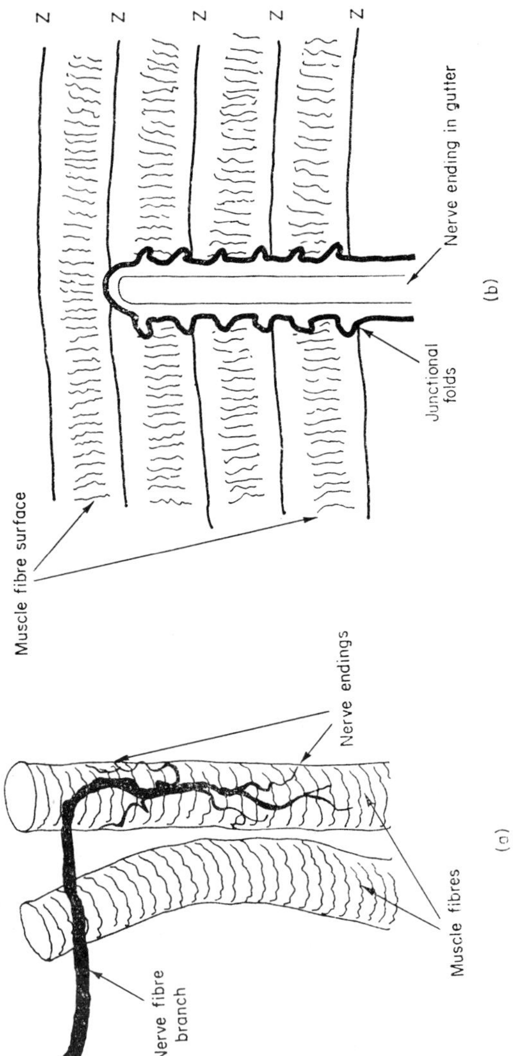

Figure 5.3. Diagram to show motor end-plate structure. In (*a*) a nerve fibre branch gives off fine nerve terminals, when it reaches one of the muscle fibres of a motor-unit. These terminals are surrounded by numbers of nuclei (not shown) which are part of the end-plate structure. In (*b*) the nerve terminal is shown resting in its 'gutter' on the muscle fibre, where the plasmalemma of the

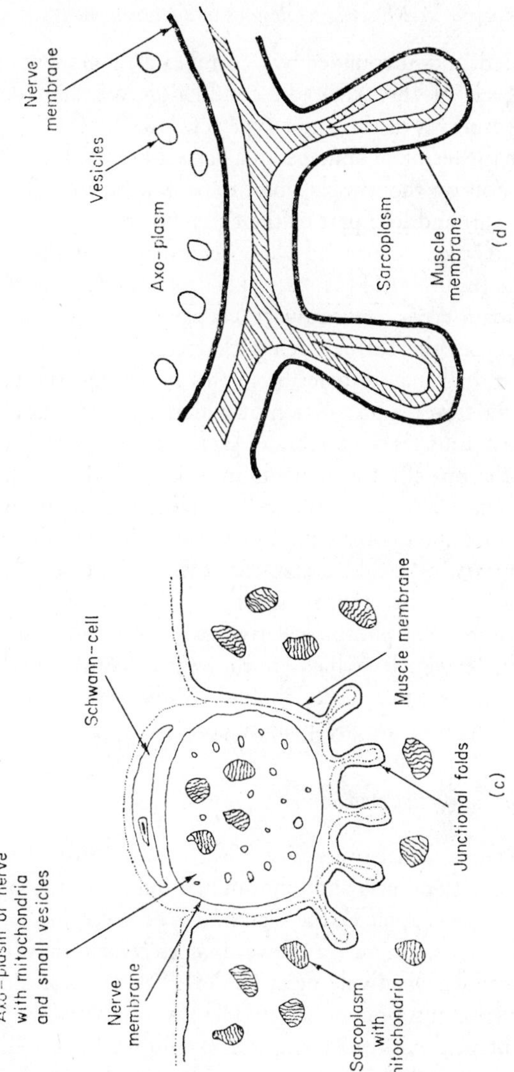

latter shows so-called junctional folds. In (*c*), a cross-section of a terminal is shown. Note that the axoplasm of the nerve lies in the gutter and is still protected on its outer surface by a Schwann-cell; also note the folds in the plasmolemma of the muscle fibre. The structure of the junctional folds is shown in more detail in (*d*). After [15], [27] and [28].

myelinated, and surrounded by a complex arrangement of Schwann cells. Much of the work on conduction we shall describe has employed nerves of this latter type (cf. [164]).

When the terminal branches of a nerve fibre reach the motor end-plate region on the muscle fibre, they split up into finer branches (figure 5.3*a*) and lose part of their myelin sheath as they enter the gutters of the motor end-plate region, where they appear to terminate (figure 5.3*b*) [15, 28, 95]. The motor end-plate regions are of two main types, the compact 'en plaque' and the spreading 'en grappe' type, the former being characteristic of the fast, twitch fibres we shall mainly describe in this book, and the latter of the slow, tonic fibres, a proportion of which also occur in many of the major 'fast' muscles of vertebrates [27]. The 'en plaque' type of end-plate is the one illustrated in figure 5.3 (*a–d*). Note that its nerve terminals have a triple-layered cell-membrane of their own, in close contact with the plasmalemma of the muscle fibre they serve; the latter appears to be folded inwards at the point of contact. Within the nerve terminal we can see mitochondria which supply the energy source via phosphorylative oxidation, and also numerous other smaller vesicles which are thought to contain packets of the chemical transmitter, acetylcholine, ready for release when the impulse arrives in the end-plate region [15, 95].

Membrane and action potentials

Apart from the discontinuity at the motor end-plate, and again between the transverse and longitudinal tubules of the sarcoplasmic reticulum (see chapter 4), the membranes of nerve and muscle fibres behave in such a similar way towards an electric impulse that in describing one we effectively describe the other, save for differences in the time constants of conduction. Here we shall mainly rely on the results obtained with the giant non-myelineated nerve fibres of the squid, Loligo, and to some extent with the smaller, myelinated fibres of the frog.

Conduction along the membranes of nerves and muscles depends upon the fact, common to all living cells, that there is an electrical potential difference between the outside and the inside of the cell

[76]. Thus, the resting cell is said to be *polarized*, being charged negatively on the inside and positively on the outside. This potential difference, whether in plant or animal cells, is due in the overwhelming majority of cases to the effect of a high internal concentration of K and protein ions and a high external one of Na and Cl ions, and to the fact that the resting membrane is at least ten times more permeable to K^+ and Cl^- than to Na^{++}. There is thus a tendency for K^+ to escape from inside, and for Cl^- to enter from outside much faster than Na^+ can. It is this which sets up the observed potential difference.

The resting cell membrane may be treated either as a Cl^- or a K^+ electrode, the latter being the usual convention adopted. Then in common with all types of electrode, the potential it can generate is given by the Nernst formula:

$$V_K = (RT/\mathscr{F}) \ln [K^+]_o/[K^+]_i = 58 \log_{10}[K^+]_o/[K^+]_i \qquad 5.1$$

(at 20°C)

where V_K is the equilibrium potassium potential = potential difference between outside and inside; R is the gas constant, T the absolute temperature, and $\mathscr{F}$ the Faraday: $[K^+]_o$ is the outside, and $[K^+]_i$ the inside, $[K^+]$.

Using the ratios of the K^+ concentrations given in table 5.1, we find that the resting potential of the squid axon should be −74 mv (inside relative to outside), whereas that of the frog muscle fibre will be considerably higher, about −97 mv, if the only factor of importance is the relation of external to internal $[K^+]$. Table 5.2 shows that both these potentials are higher than those observed, probably due to damage to the excised tissues [76]. Thus the idea that the resting cell membrane acts as a pure K^+ electrode is substantially correct. In support of this, we may calculate what the resting potential would be if exactly the opposite situation obtained and the membrane were more permeable to Na^+ than to K^+; the value now becomes +55 mv for the squid axon, inside to outside, instead of the −70 mv or so, observed for the resting potential. As we shall see, this is in fact the right sign and almost exactly the right magnitude for the final potential, observed when the membrane becomes depolarized by an electrical impulse, and the so-called action potential spike is propagated. The total potential change would be +125 mv.

Measurement of resting and action potentials in large nerves such as those of the squid can be made by first inserting a fine glass capillary, filled with saturated KCl solution, into the cut end of the nerve and pushing it 20 to 30 mm down into the axoplasm (see figure 5.4*a*) [76, 95]. This forms the inner electrode, the outer also being a KCl electrode, resting on the surface of the nerve. With smaller nerves, a fine capillary with a tip diameter of about 1 μ can be inserted transversely through the nerve membrane to form the inner electrode. In either case, the potential generated between the inner and outer electrodes is fed into an amplifier circuit and displayed on the tube of a cathode-ray oscillograph. When the internal electrode is inserted, it becomes immediately negative with respect to the outer electrode, to the extent of the 70 mv or so, characteristic of the resting potential. Note that in all further discussion, the potential difference will always be given in the sense of *the inside relative to the outside* of the cells or fibres.

Now what happens when we stimulate the nerve with a single electrical impulse, some distance away from the recording electrodes (see figure 5.4*a*)? We find that after a very brief delay, due to the time taken to conduct the impulse to the recording electrodes, there occurs a rapid swing of the potential difference to 40–50 mv positive, followed by a somewhat slower decay back to, or slightly below, the original resting potential of −70 mv. As seen from figure 5.4*b*, the whole process in squid nerve is over after about 6 millisec at 16°C. It is known as the propagated action potential (AP), which we can regard as a wave of depolarization/repolarization passing in either direction along the nerve or muscle fibre from the point of stimulation [76]. To illustrate what is meant by propagation of the AP, let us consider a hypothetical nerve of great length, with a conduction velocity of 25 metres/sec: we record the AP with one set of electrodes 2·5 cm away from the point of stimulation, and another set 25 cm away; we will then observe two AP spikes of the form shown in figure 5.4*b*, both identical in shape and height, but separated in time from the stimulus by 1 millisec and 10 millisec, respectively.

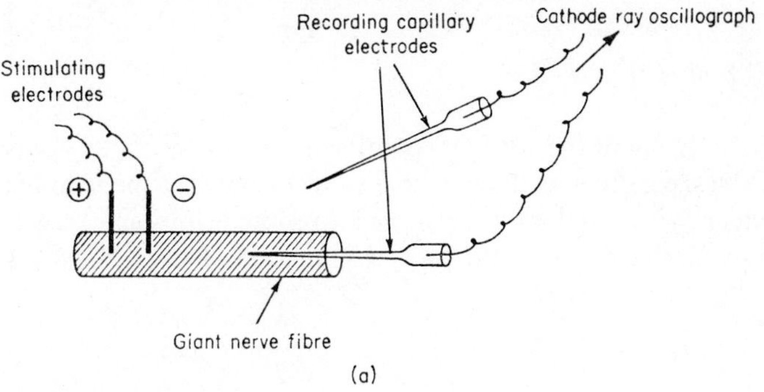

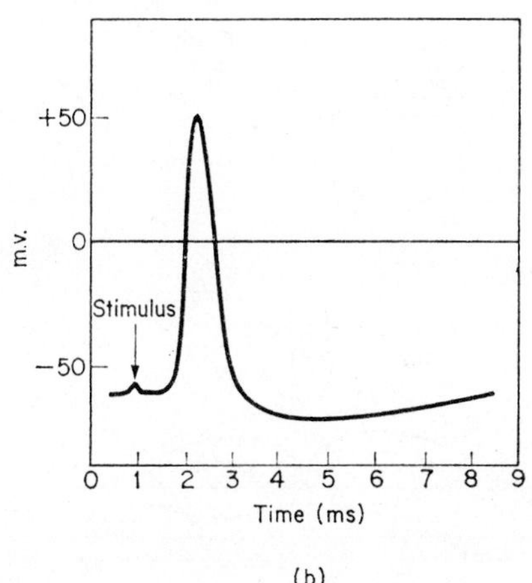

Figure 5.4 (*a*) Diagram to illustrate the method of recording resting and action potentials in giant nerve fibres (axons) such as those in the squid.
(*b*) The form of the action potential spike after a single stimulus delivered to a squid axon in sea-water at 18°C. After [76].

Mechanism of conduction

The swing of the AP to its positive maximum of about 50 mv is of the same order as the potential to be expected if the membrane suddenly became locally permeable to sodium ions and relatively impermeable to potassium (+55 mv): in other words, it appears to

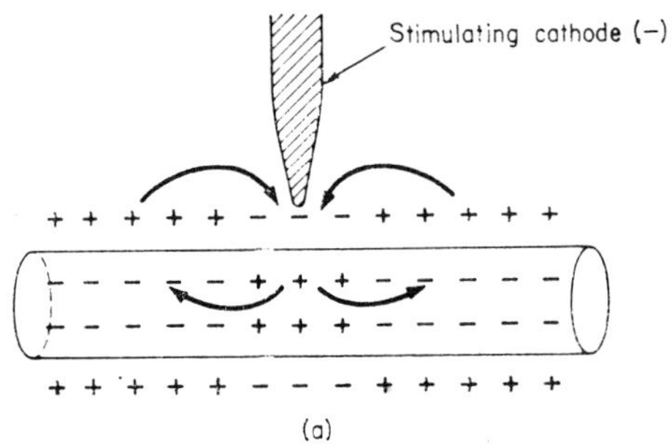

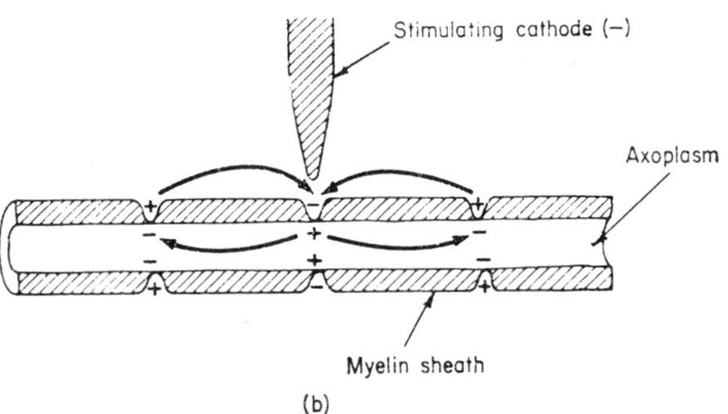

Figure 5.5. (a) The way in which the action potential is propagated away in both directions from under the stimulating cathode, in a non-myelinated nerve fibre. (*b*) Saltatory conduction in a myelinated nerve fibre. Note that the action potential spike jumps, as it were, across the insulated internode region, without having to effect individual ionic changes in between. The action potential originates at the nodes. There is a flow of current both inside and outside the internode region, the latter occurring in the bathing, extracellular fluid. After [76].

behave in the active state as a sodium electrode, instead of a potassium electrode [76]. Accepting for the moment that this is indeed the mechanism of the generation of the AP, we can then think of the train of events as shown in figure 5.5*a* and *b*. The resting parts of the membrane are charged negatively inside and positively outside. However, at the stimulating cathode (—), we have suddenly imposed a negative potential from without, so that immediately under this electrode the charge is reversed, and the membrane suddenly becomes permeable to Na ions. This generates a new local potential difference (inside to out) of +55 mv. This local depolarized region is, of course, electrically in contact with the regions on either side of it, so that there too current begins to flow, carried by the inward flow of Na ions down their concentration gradient, in exactly the same manner as happened locally under the stimulating cathode. Thus the AP is conducted away in either direction from the point of stimulation in the form of a local inward flux of Na ions. Note that while this influx is taking place, the membrane becomes temporarily impermeable to K ions.

How do we explain the descending limb of the AP, that is the repolarization wave? This is comparatively simple, because the permeability to K ions is restored as the sodium current begins to fall off. Then, due to the decline in sodium permeability immediately after the stimulus has passed, there will be excess positive charge on the inside of the membrane: this has the effect of driving K ions out and down their concentration gradient in equivalent numbers to the Na ions which have entered, thus neutralizing the charge and repolarizing the membrane. This so-called K-current is generated much more slowly than the preceding Na current, but eventually tends to overshoot it, as we see from figure 5.6. This may be due to a small amount of Cl flux during the passage of the APin any event; it is soon counterbalanced, and the resting potential restored to its original value.

We have no space here to describe the elegant experiments which have been done to show the time course of the sodium and potassium currents [77]. Suffice it to say, that this depended on the use of a 'voltage' clamp, whereby a constant voltage was superimposed on the resting potential, to give the desired voltage difference across the membrane. It is then possible to measure the changes in the sodium and potassium conductances with time after the stimulus and against varying initial membrane potentials. The form of the conductance

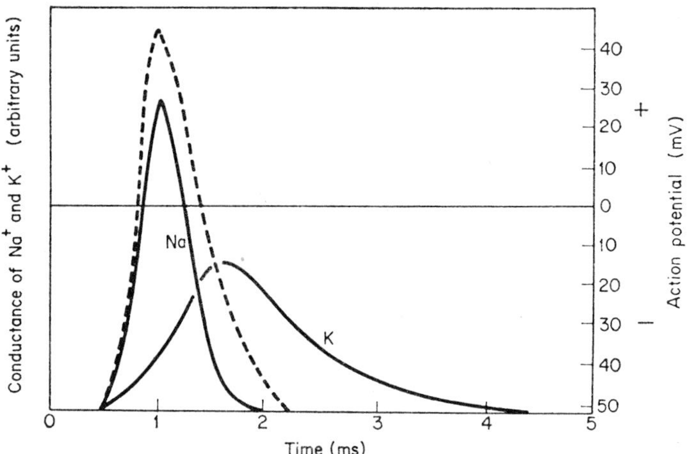

Figure 5.6. **Diagram to illustrate how the sodium and potassium conductance of a nerve fibre changes as an action potential passes the recording electrodes (full lines). The form of the action potential is given by the dotted lines.**

changes, with the membrane at its normal resting potential of −70 mv, is shown in figure 5.6. We note the much slower increase of the K conductance, and how this explains the form of the AP. In this case, the quantity of K ions passing out exactly balanced that of the Na ions which had entered (about 4·3 pmole/cm² of membrane).

Quantitative and metabolic aspects

We saw above that the total transfer of Na and K ions during the passage of one AP in squid nerve amounted to only about 4 pmole/cm² (1 pmole, or picomole = 10^{-12} gram ion). The total $[Na^+]$ and $[K^+]$ are each of the order of 400 mmolar = 4×10^{11} pmolar, and the volume corresponding to 1 cm² of membrane of squid nerve (diameter 0·05 cm) is about 0·00625 cm³. A total of

$2{\cdot}5 \times 10^6$ pmoles of these ions is thus available in this volume, so that only about 1 millionth of the total number is transferred per impulse. It follows that the metabolic energy required to restore the *status quo* is minute, even after a long train of impulses; for this reason, restoration can be delayed for a considerable time without much damage being done [76].

Compared with the very rapid, but quantitatively insignificant transfer of ions across the membrane during the passage of an impulse, the rate of sodium and potassium flux in the resting state is very low indeed (about $\frac{1}{2000}$ of the rate during the AP, in squid nerve). Nevertheless to establish the large ionic concentration differences which exist in the resting state, a metabolic pump of some sort must be operating, and this pump will have to work continuously throughout life to counteract the inevitable Na influx and K outflux, which is due partly to the very slow passive diffusion of Na and partly to the effect of numberless impulses. The pump which exists for this purpose in most nerves and muscles is an Na pump, using the energy from ATP hydrolysis to drive out Na ions against the concentration gradient, and being refuelled from oxidative phosphorylation carried out by localized mitochondria; no K pump seems to be necessary here, because of the high K permeability of the membrane. The operation of the Na pump in the resting state, and also immediately after a train of impulses, has been elegantly demonstrated by efflux experiments, using ^{24}Na [76, 95].

That the fundamental process of conduction is quite independent of metabolism and of the Na pump has been shown up, not only by flux studies, but by the direct experiment of extruding axoplasm from squid nerve [9]. It turns out that axoplasm can be pushed or 'rolled' out of these nerves without damaging the membrane in any way, by employing a miniature 'lawn roller'. It can then be shown that the portion of the nerve which is empty will not conduct an impulse when bathed in sea-water, unless axoplasm from an adjoining segment is pushed into it, or the empty space is filled with an isotonic solution of K ions. A completely artificial internal environment can be substituted for the axoplasm, the main requirement being the correct concentrations of K ions inside and Na ions outside, whereas the type of anion added is optional within wide limits.

Such a system gives normal action potentials which it will propagate for many hours. Moreover, it has the immense advantage that the effect of varying internal Na and K concentrations can be studied, both on resting and action potentials. When this is done, it is remarkable how closely the results agree with the expectations of the Nernst equation (5.1) in its simple form, that is, the membrane acts as a K electrode in the resting state, and a Na electrode when the AP starts off.

These 'lawn-roller' experiments incidentally dispose of the idea, propounded in old histology text-books, that the nerve fibre contains finer fibrils within it. The 'fibrils' are evidently artefacts of the fixation technique. The experiments also argue against any chemical transmitter, such as acetylcholine, being involved in propagation of the nerve AP, because such a substance would almost certainly have been washed out, along with the axoplasm, during the course of preparation.

Conduction rate in nerves and muscle

The fundamental process which propagates the action potential is common to nerve and muscle, so that the action potential spikes themselves and their duration are also of similar magnitude in both sorts of tissue. Large differences exist, however, in the conduction velocity, as we see from table 5.2 [41]. This shows first that myelinated nerves conduct much faster, at comparable diameters, than do non-myelinated nerves. This is due to the so-called saltatory conduction of myelinated fibres, whereby the AP can jump from node to node over the highly insulated internode region, rather than having to traverse it (see figures 5.2*b* and 5.5*b*). Were it not for this adaptation, the nerves of mammals, for instance, would have had to grow very thick indeed to conduct at the velocities observed. This is because the rate of propagation depends on the square root of the diameter in non-myelinated nerves, but on the first power of the diameter in myelinated.

We also see from table 5.2 that the conduction rate in the muscle of the frog is much lower than in the myelinated motor nerves which supply it. This is because the muscle cell membrane is more or less uninsulated, and behaves rather like the membrane of a non-myelinated nerve. There are similar differences in the so-called absolute refractory period. This is the short period immediately after a single AP, when the nerve or muscle will absolutely *not* conduct an impulse; it has about the same duration as the AP itself, and is followed by a relative refractory period, during which the stimulus strength has to be increased above threshold to start off an AP [76, 95]. We should note that this threshold strength is always below the value of the resting potential, and is of the order of 10–20 mv. It is logical to assume that it is the potential at which the inward Na current induced by the stimulus just balances the outward K current; the theoretical potential would be about 19 mv.

The question of conduction velocity leads us immediately to the problem of what happens at the motor end-plate, the first membrane discontinuity, and also later at the junction of the transverse and longitudinal tubules of the reticulum within the muscle fibre, which is the second discontinuity.

Transmission across the motor end-plate

The events which occur at the motor end-plate, when a single AP spike passes down the motor nerve and arrives at the nerve terminal, can be qualitatively described as follows: acetylcholine is released from the small vesicles in the nerve terminal, diffuses across the minute space between it and the end-plate membrane, in contact with the muscle membrane, and immediately depolarizes this membrane (see figure 5.3) [15, 95]. At almost exactly the same time as the depolarization occurs, the acetylcholine is destroyed by an extremely rapidly acting choline-esterase which is present in the end-plate region, so that the depolarizing effect does not last very long. The depolarization is of a different kind from that which occurs in nerve and muscle membranes when they are stimulated, the end-plate membrane in this case becoming unselectively permeable to both Na and K ions, instead of to Na ions only, as in the latter case. This means that the end plate potential (epp) can never exceed the resting potential. Nevertheless, this change of potential at the end-plate will automatically induce a local change of potential, of about the same size, in the muscle membrane in close contact with it, and this is sufficient to bring about local depolarization, and the inward flow of Na ions. A propagated AP starts off from this depolarized region of the muscle membrane, in exactly the manner we described for propagation in nerve fibres.

There are many interesting quantitative features of the epp, which we have no space to discuss here in detail. We may mention, however, the effect of curare which somehow blocks transmission along the muscle membrane. Nevertheless it leaves unaffected the release of acetylcholine at the nerve terminal, and thus allows the development of a normal epp, which slowly dies away in either direction along the muscle fibre, without propagating an AP [76, 95].

Transmission by the sarcoplasmic reticulum and initiation of the muscle twitch

So far, we have been able to describe the events which occur in stimulated nerves, end-plates and muscle membranes in quantitative electrical and chemical terms. However, we can no longer do this when we come to discuss the question of transverse and longitudinal transmission of the impulse within the muscle fibre itself, because it has not yet proved possible to record the potential changes which occur in the transverse tubules and at the triad junctions of the sarcoplasmic reticular system (cf. figure 5.1 and plate IV.2). We can, however, be fairly sure that the AP transmitted inwards along the transverse invaginations of the plasmolemma, either at the A–I junctions in mammals or the Z-lines in frogs, for instance, is of the same amplitude as the 'parent' AP, but probably much slower, because the diameter of the tubules is so small. From that point onwards we have no exact idea of the chain of events, except that the final result is the release of Ca ions, probably from their storehouse in the TC of the triads and the longitudinal vesicles of each fibril, their diffusion to the active enzyme sites on the actin and myosin filaments, and their immediate 'explosive' effect in activating the

ATP-ase sites. This latter process initiates movements of the myosin heads and contraction, by sliding of the two sorts of filaments over one another.

The proof of the transverse inward transmission of the AP rests first on a classical experiment, done several years ago, on the effect of very local stimulation of the muscle membrane [85]; and secondly on the unravelling of the complexities of the reticular structure and particularly on the demonstration, by the use of ferritin particles, of the continuity of its transverse elements with the plasmalemma (see plate IV.1) [49, 84].

The stimulation experiment, which has to be seen on film to be fully appreciated, can be crudely illustrated as in figure 5.7, where

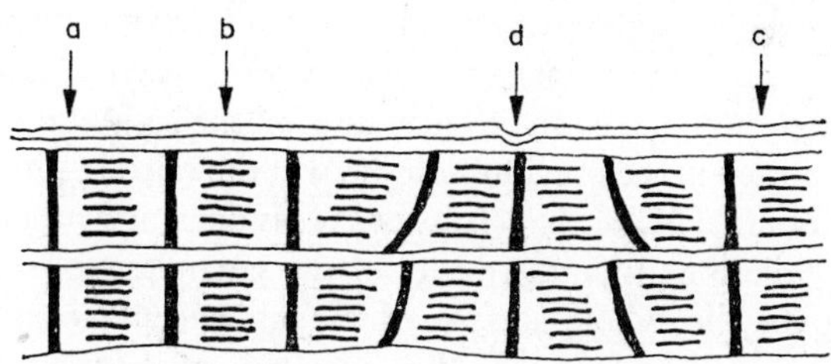

Figure 5.7. Diagram to illustrate the effect of stimulating a single frog muscle fibre with a micro-electrode at various positions along the sarcomeres [85]. Note that the stimulus is effective only at a Z-line (*d*), and brings about very localized shortening of the stimulated half sarcomeres, and not the others on either side of them. The contraction penetrates only a few fibrils in depth. Stimulation in the I-band (*a*), A-band (*b*) or A–I junction (*c*) is completely without effect.

a micro-electrode is moved slowly along the surface of a single frog muscle fibre, and a tiny impulse delivered from it at the positions shown. The impulse is so small that it can only activate the first few fibrils under the surface: as we see, it will only induce a local contraction when it is placed exactly over a Z-line. At intermediate positions, nothing happens at all. Now in frog muscle, the Z-line is the position at which the transverse tubules of the reticulum invaginate the fibre, and this combined with the knowledge that these tubules are continuous with the outer membrane, is sound proof that this is the sole mode of inward transmission of the impulse.

This demonstration at once clears up the long disputed question of how an impulse could arrive at the centre of a thick muscle fibre (say, 100 microns in diameter) quickly enough to give the observed speed of contractile response. Quite clearly simple diffusion is out of the question, but if the impulse is really a propagated AP, the difficulties are removed at one stroke.

According to our description of inward transmission, the impulse has now arrived at the triad junctions of the fibrils within the muscle fibre. How it is transmitted across the gap in the triad system is unknown; the release of acetylcholine is one possible mechanism, but there is as yet no evidence of its participation. Certainly the need for a chemical transmitter at the junction would seem to follow from what occurs at the motor end-plate, and also in other synaptic regions in the central and peripheral nervous system. Nor do we know for certain in the intact reticular system, where calcium ions are stored, although we can be fairly sure, on the basis of the *in-vitro* studies reported in the previous section, that the calcium pump itself is situated in the TC of the triads [66]. This does not mean, however, that the longitudinal vesicles of the sarcoplasmic reticulum, which cover the whole surface of each fibril, are incapable of *storing* calcium, but only that they may not actively accumulate it.

The direct proof that Ca is released when the AP finally arrives at the triad region of the fibrils has been demonstrated recently, by an elegant photometric technique [90]. The basis of the method is to feed toads on murexide, and at the same time to inject methyl sulphexide. The latter compound is known to increase the permeability of membranes to large molecules, and so the murexide first goes into the animal's blood stream, and thence is ingested into the sarcoplasm of its muscles. Since murexide changes colour when it binds Ca ions, the release of these ions into the sarcoplasm after a stimulus can be detected by simultaneous photometric recording at two wavelengths, and subsequent analysis of the difference spectrum. The type of curve obtained is shown in figures 6.2 and 7.2. We notice two important features of the process: first, that Ca release begins some time after the passage of the AP (first signal), and is followed immediately by the onset of contraction; second, the

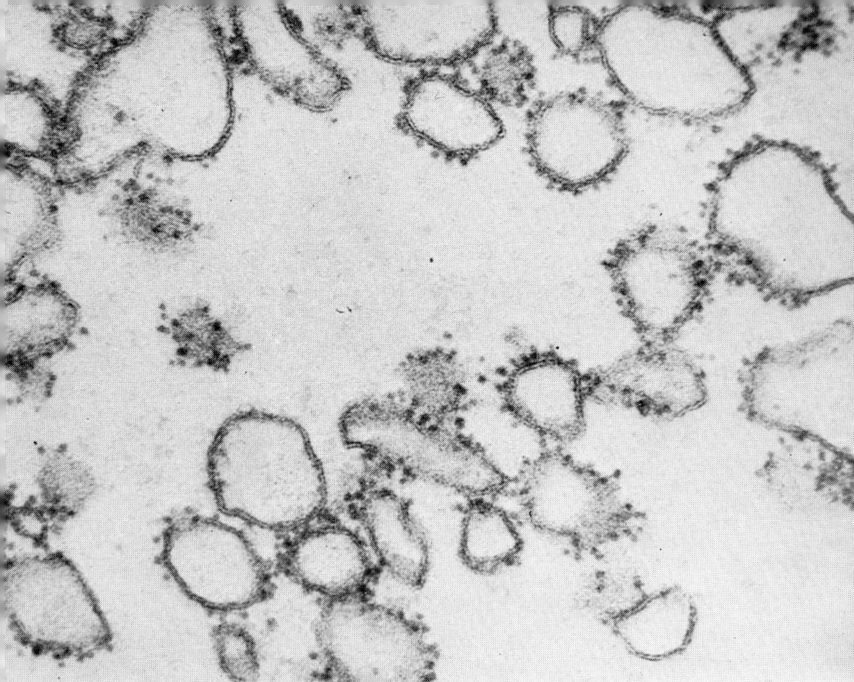

Plate IV.3a Sarcoplasmic reticular fragments, after treatment with alkali, followed by Hg-phenylazoferritin. Note how the SH-reagent has become attached round the periphery of the particles. In many places, the triple layered structure of the membrane can be seen. These particles almost certainly arise from the longitudinal swellings of the triads, and the vesicles shown in plate IV.1(*a*) and (*d*).

Plate IV.3b Vesicles treated with 'ferritin', as above, but without alkali. In this case the whole surface has become covered in 'ferritin' granules, often oriented in rows. (Photos by courtesy of Dr W. Hasselbach.)

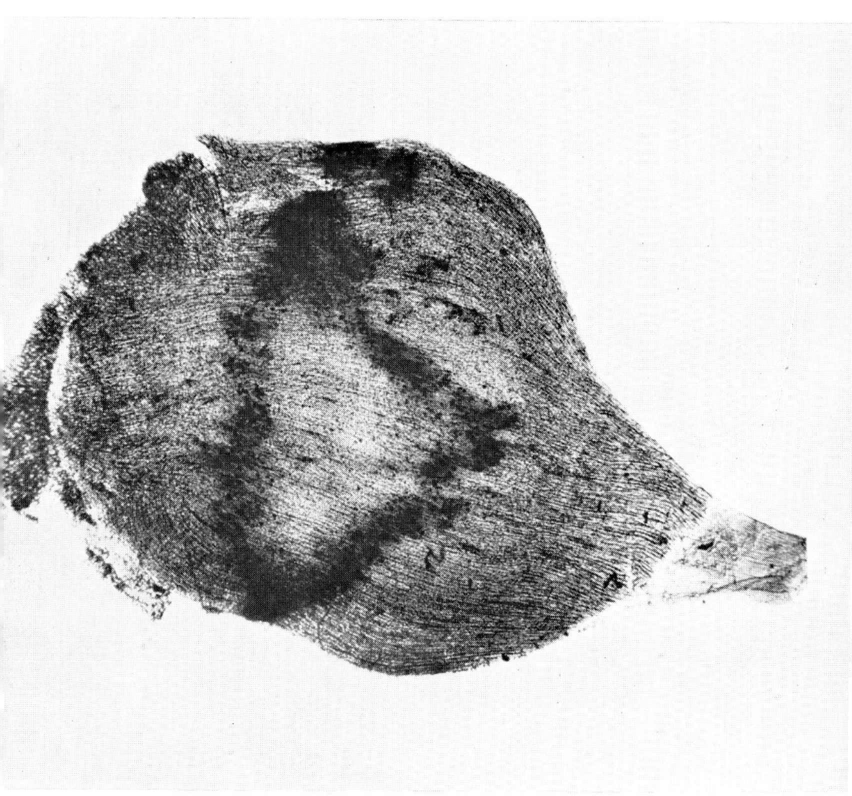

Plate V. Mouse muscle (biceps brachii) squashed between glass slides, and then stained for choline-esterase in the end-plate regions with acetylthiocholine iodide. Note how the end-plates are situated towards the centre of the muscle, in a ring.
(Photo by courtesy of Dr G. Goldspink.)

contraction, a single twitch in this case, has hardly reached two thirds of its maximum, before the calcium 'wave' seems to have completely disappeared again (figure 7.2). There is no further change during the comparatively slow phase of relaxation.

How do we explain this calcium flux, which at first sight appears so anomalous? We can assume that the first half of the wave is due to rapid release of Ca from within the Ca pump in the TC of the triads; in other words the permeability of the TC membrane to Ca is suddenly changed by the AP, in a manner analogous to the changes in sodium permeability along the outer plasmalemma of the fibre, when an impulse passes.

We saw in chapter 3, that two Ca ions are needed per enzyme site on the myosin filaments, to set the contractile ATP-ase in motion, and since this is the case, the total amount of Ca required to saturate all the sites would be about $2{\cdot}4 \times 10^{-4}$ moles per kg muscle. Presumably this is also the total amount released by the TC of the triads during a single twitch. A sudden release of such a quantity into the sarcoplasm easily explains how murexide is able to pick up some of it and change colour, before the Ca^{++} arrives at the enzyme sites. Similarly, if the binding constant of Ca^{++} to myosin or tropomyosin is, say, one order higher than that to murexide, it is easy to explain the second downward half of the Ca^{++} wave, during which Ca^{++} seems to disappear from the sarcoplasm, before the twitch has reached its peak: the active sites will have adsorbed most of it.

The situation during the return journey to the TC of the triads in the relaxation phase is, of course, quite different. The first step must be repolarization of the triad membrane, and the onset of vigorous pumping activity, because the free Ca^{++} concentration in the sarcoplasm, although probably not greater than 10^{-6} M, is still sufficiently high to stimulate such pumping (cf. figure 4.3). Hence, Ca ions are slowly dragged away from the myosin sites, and back into the TC. Because the process, like relaxation itself, is comparatively slow, the free concentration of Ca^{++} in the sarcoplasm will remain low, and probably insufficient to be detected by the murexide reaction. Hence the apparent form of the Ca^{++} wave: first a rapid release of Ca^{++}, partly picked up by murexide; then grabbing of this Ca^{++} by the active sites; and finally, renewed pumping back of Ca^{++} into the TC of the triads, at low concentration.

This description of the process of activation although, satisfactory in qualitative terms, leaves very much to be desired from the point of view of a quantitative reaction scheme. The chief hiatus is the mechanism of the sudden Ca release from the TC. Is this to be explained by depolarization of the membrane, in the same way as sodium influx at the plasmolemma, or is another mechanism needed? We can at least be sure of one thing from the *in-vitro* studies: mere inhibition of the Ca pump would not account for the facts, because under such conditions Ca efflux from triad fragments is not sufficiently rapid. Hence, we seem to be left only with the depolarization/repolarization hypothesis, and in that case we need to know the resting potential difference across the TC membrane, due to the Ca concentration gradient between inside and outside. If this gradient is about 500/1, as *in-vitro* studies suggest, and the membrane is moderately permeable to Ca ions, at least while the pump is operating at high ATP concentrations, then the TC membrane potential would be +157 mv, inside to outside. Although this is a very high potential, it is of the correct sign, in relation to the contents of the fibril. However, we do not know what counter-ions are involved, nor the nature of the anion required to balance such a Ca^{++} gradient, although inorganic phosphate is the most likely candidate. Hence, the major questions to be resolved are: first, the mechanism of transmission and of Ca^{++} release on arrival of the AP at the transverse tubule–TC interfaces, and secondly, the nature of the other anions and cations, involved in the pumping of Ca^{++} *in vivo*. As we saw earlier, Mg^{++} is certainly required, and inorganic phosphate is always present in sufficient quantity to act as the anion ($\sim 5 \times 10^{-3}$ M, even in resting muscle [31, 63].)

Table 5.1. Distribution of electrolytes on either side of nerve and muscle membranes [41]

Tissue	*Concentrations inside (in mM)*			*Concentrations outside (in mM)*			*Ratios inside to outside*		
	Na	*K*	*Cl*	*Na*	*K*	*Cl*	*Na*	*K*	*Cl*
Squid nerve	49	410	40	440	22	560	1 : 9	19 : 1	1 : 14
Sepia nerve	43	360	—	450	17	540	1 : 10	21 : 1	—
Crab: leg nerve	52	410	26	510	12	540	1 : 10	34 : 1	1 : 21
Frog: sartorius muscle	15	125	1·2	110	2·6	77	1 : 7	48 : 1	1 : 64

Table 5.2 Conduction rates in nerve and muscle [41]

Tissue	*Temp.* (°C)	*Myelinated* (M) *or non-myelinated* (N)	*Fibre diameter* (μ)	*Velocity m/sec*	*Observed Resting potential*	*Observed Action potential*
Crab nerve	20	N	30	5	−62	+120
Squid-giant axon	20	N	500	25	−61	+89
Cat non-myelinated nerve	38	N	0·3–1·3	0·7–2·3	—	—
rog muscle fibre	20	N	60	1·6	−88	+120
rawn myelinated nerve	20	M	35	20	—	—
Frog ditto	24	M	3–16	6–32	−71	+116
Cat ditto	38	M	2–20	10–100	—	—

6: Mechanics of Contraction

We saw in the last chapter how the action potential is conveyed from the motor end-plate along the plasmolemma of the muscle fibre, and then inwards via the transverse tubules of the sarcoplasmic reticulum, where it finally activates the release of Ca ions from the TC of the triads of each fibril, and thus initiates ATP splitting and the muscle twitch. We shall now consider mechanical aspects of the twitch in detail, and show how the fusion of twitches results in a tetanus. For this purpose, the muscle is usually stimulated directly by placing massive electrodes on its surface, rather than via its nerve, because this method cuts out the delay in transmission at the motor end-plate, and is generally more convenient.

Note that when we speak of *isometric* experiments, we mean those in which muscle length is fixed and the contraction therefore develops tension in the connections, which is recorded by a strain-gauge or tension-transducer. On the other hand, *isotonic* experiments are those in which the muscle is allowed to *shorten* against a constant load (= tonos).

Experimental set-up

There are many possible set-ups which can be used for studying the twitch; one of the most generally useful and adaptable is shown in figure 6.1 (cf. reference 89). This consists primarily of a light duralamin lever, *L*, pivoted at *F*. The muscle is attached to one end of the lever via a light chain, which is tied firmly with cotton to the tendon of the muscle. The other end of the muscle, usually still attached to a piece of bone, is tied to a rigid support in a bath. The muscle rests on platinum stimulating electrodes, SE, arranged alternately negative and positive.

When isotonic twitches are to be studied, the muscle can be loaded at position *P*, on the other side of the rigidly supported pivot, *F*. When tension is to be recorded, as well, a tension transducer, *T*, is attached by a chain, at either position *A* or *B*. In position

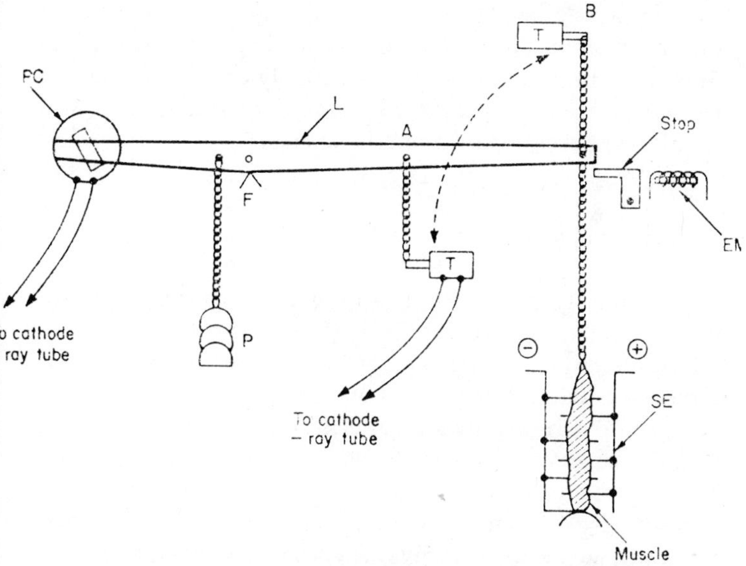

Figure 6.1. **Experimental set-up for twitch and tetanus experiments. Note the various stops for controlling the length of the muscle, and also how the tension transducer can be moved so as to measure tension either in an isotonic (A) or an isometric contraction (B). After [89].**

A, the transducer will record changes of tension during an isotonic twitch, as the muscle pulls the lever down and thereby slackens the attached chain. It can then also be used as an afterstop, by moving it up or down on its rigid stand (not shown), thus preventing the afterload, *P*, from unduly stretching the resting muscle. With the transducer in position *B*, only isometric tension will be recorded, any force developed by the contracting muscle acting directly upon the transducer, without moving it appreciably, but instead causing an increased flow of current in it, which can be displayed, after amplification, on a cathode-ray tube.

During isotonic contractions, when the afterload, *P*, is being lifted by the downward shortening of the muscle, the movements of the lever are recorded by means of the photocell, *PC*, the output from which is also displayed on a cathode-ray tube. Alternatively, the photocell can be replaced by a special type of transducer, adapted to recording large movements. Finally, the stop, *S*, on the muscle side of the pivot, can be used to hold up a contraction at any desired length, or to release an isometrically contracting muscle to any desired extent. It is controlled by the electromagnet, *EM*, which is activated by an impulse from the timing system.

Early events in the twitch

With suitably refined recording apparatus, it is possible to measure electrical and mechanical changes in the muscle within the first 1 or 2 msec after stimulation [135]. For this purpose, time marks on the cathode-ray tube are made at about 10,000 per sec, and the tension changes are measured by a piezo-electric transducer. The absolute rates of change vary considerably with the type of muscle, and with the temperature, but here we shall consider only the frog sartorius, contracting at 0°C, since these are the conditions under which many of the measurements of energy changes have also been made. To do this, a number of separate observations by various authors have been collated (references 135, 90), and corrected where necessary for temperature effects, using a temperature coefficient ($Q_{10°C}$) of 3·0 [3, 67]. Unfortunately, data for Ca release in frog sartorius are not yet available, so that the curve in figure 6.2 has been extrapolated from the data for the slower muscles of toads (see figure 7.2), and can only be said to give a general impression of the change.

Figure 6.2 shows the changes which occur in the early stages of an isometric twitch, during and just after the so-called latent period, before the muscle has developed any positive tension. We see how the action potential (AP) starts off from the moment of stimulation, has reached its peak in about 1·5 msec, and has decayed to less than half its maximal value before any other change can be detected. At 7 msec after the shock, however, the muscle tension suddenly drops and reaches a minimum at about 11·5 msec. It then abruptly reverses sign, and after another 3·5 msec, has regained its initial value, building up rapidly from that moment onwards to the full isometric

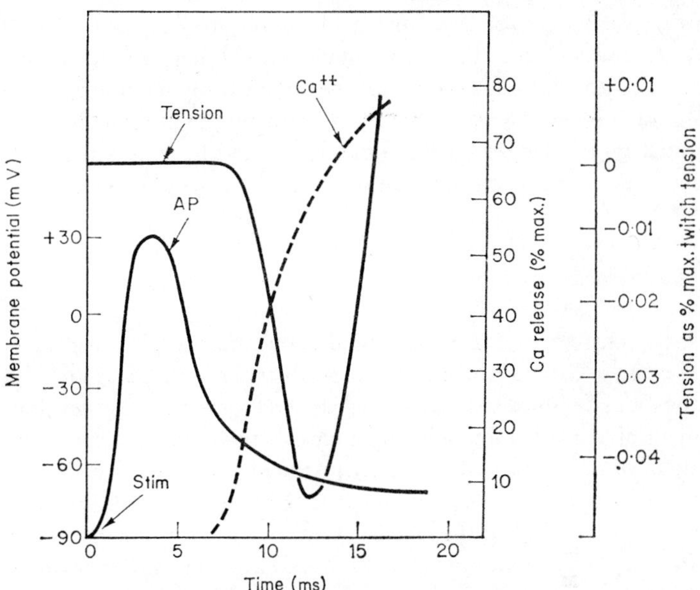

Figure 6.2. **Diagram to show the early stages of an isometric twitch. Note how the action potential has nearly died away before latency relaxation (LR) and the release of Ca ions has begun, and how the LR is succeeded by positive development of tension after about 12 msec. (Frog sartorius muscle at 0°C.) After [135,] [137] and [90].**

twitch tension (not shown). This early drop of tension is known as *latency relaxation* (LR) [135, 136]. It is minute in terms of the full isometric tension, being perhaps $\frac{1}{2000}$ of it, so extremely delicate methods are needed to detect it. It very probably coincides with the initiation of Ca release from the TC of the triads, which would be expected to start at about 7 msec in frog muscle, and will have reached 60 to 70 per cent of its maximum at the end of the latent period, judging by the results with toad muscles.

An explanation of the LR has recently been advanced in terms of length changes in the sarcoplasmic reticular system [137]. The triads and longitudinal vesicles of this system enwrap the fibril closely and probably make up at least 15 per cent of the muscle volume. Since, it is assumed, they rapidly lose their store of Ca ions during the passage of the AP, this might be expected to cause a considerable decrease in their internal osmotic pressure, and hence a flow of water from them. The tubules would then tend to stretch, and this stretch could account for the minute drop of tension during the LR period. The concept is supported by the known effects of initial muscle length on the duration of the LR, stretched muscle showing a longer latency than unstretched, as would be expected.

Isometric and isotonic twitches

The changes in tension discussed above are complete within 15 msec at 0°C, and are much too small to be noticed at all with apparatus designed to measure the later and more dramatic development of active tension. The full course of this development during an *isometric twitch* of frog sartorius muscle is shown by the top curve of figure 6.3 [89]. We see that active tension builds up very rapidly at the end of the latent period, and reaches a maximum in about 170 msec after the shock. From 200 msec onwards it begins to decline again, increasingly rapidly at first, and then after about 450 msec, along an approximately exponential time course. Even at 900 msec, however, there is still some active tension in the muscle. This ability of the muscle to bear tension can only be due to active physical and chemical processes going on within it, and so in a sense, the decay of tension is a measure of the decay of these processes, or in other words, of the decay of the so-called *active state* [89, 68].

The isotonic twitch contrasts vividly with the isometric one, as we see from figure 6.3 which shows the length changes which occur during a series of isotonic twitches of the same muscle under varying loads [89]. We note first that shortening during the isotonic twitch only begins when enough tension has been developed by the muscle to equal that in the afterload; hence the twitch appears to start off later and later the larger the afterload. Shortening, and with it lifting of the load, begin along a nearly linear time course, but reach their maxima earlier and earlier the larger the load: for instance, after about 375 msec under a load of 3 g, 340 msec under 5 g and less than 300 msec under 9 g. Relaxation then sets in with increasing

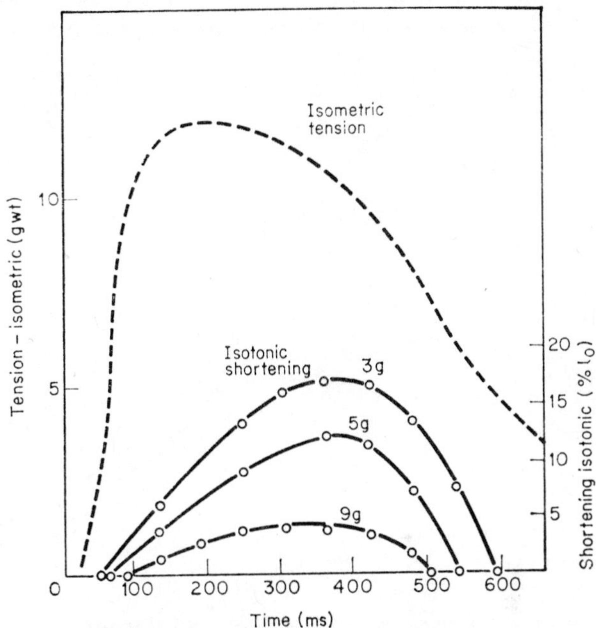

Figure 6.3. **The form of the isometric and isotonic twitch (frog sartorius at 0°C). Note how the isotonic twitches, under various loads, have died away several hundreds of millisecs before the tension has declined to zero. After [89].**

velocity, and like the peak shortening, is also completed earlier the larger the load. Even when the muscle has relaxed back to its initial length, there is still considerable tension left in it even under the lightest load, as we see by comparing the curve for a 3 g load with that for the isometric twitch. Such a result demonstrates how careful one must be in interpreting the results of isotonic twitches, since it is obvious that mere completion of the length changes is no criterion of the potential activity of the muscle at this moment. In fact, isotonic twitches which do not also include a measure of tension changes provide virtually no information on the decay of the active state.

The development and decay of tension during an *isotonic* twitch can be measured with the apparatus shown in figure 6.1, by attaching the transducer in position A (for a description of the method see reference 89). The result of this manœuvre is shown in figure 6.4,

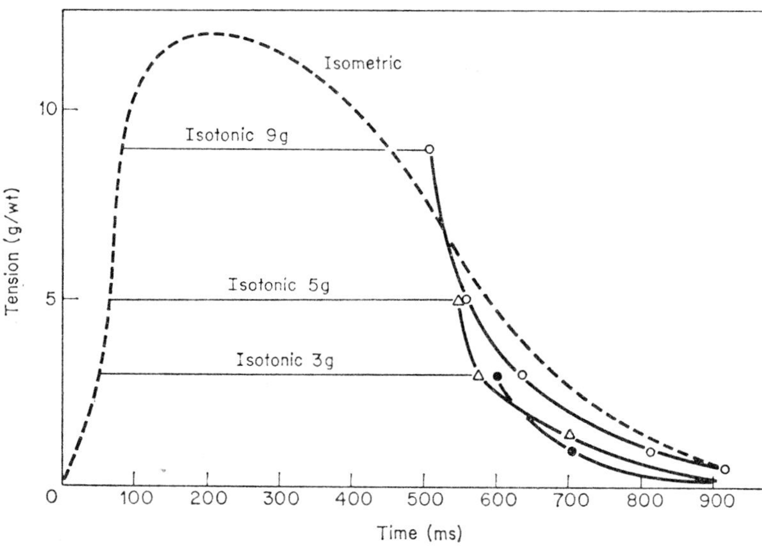

Figure 6.4. **The development and decline of tension in the isotonic twitches shown in figure 6.3, measured by putting the transducer in figure 6.1 in position A. An isometric twitch is included for comparison. After [89].**

where an isometric twitch is again included for comparison. We see that the initial build-up of tension in the isotonic twitches follows the same time course as that of the isometric twitch. In each example, the muscle begins to shorten as soon as the tension has built up to that of the afterload it is bearing, and the tension then remains constant at this value until relaxation is just complete. During the relaxation phase, of course, the afterload pulls the muscle back more or less slowly to its initial length, fixed by the afterload stop. The tension, measured instantaneously at this moment (at 500 to 600 msec), is still that of the afterload, but it begins to fall very rapidly from then onwards, although in no case does it reach zero even by 900 msec. This is a time so late in the course of an isotonic twitch that all activity would have been deemed to have ended long ago, if judged by length changes alone (cf. figure 6.3).

Another interesting feature of figure 6.4 is the difference in the time relations of tension decay, resulting from the isotonic conditions of contraction. Only the tension decay for the highest load is in broad general agreement with that for the isometric twitch, whereas at the two lower loads, where a great deal more shortening occurred during the initial phase of the twitch, the time course of decay differs quite markedly. From this, it seems that the lower the load, and therefore the greater the shortening of the muscle during the rising phase of the twitch, the sooner the tension begins to decay during relaxation. This also implies an earlier decay of the *active state*, if the latter is defined solely in terms of the tension the muscle can bear at any moment.

The effect of length changes on tension decay can be more easily demonstrated by the method of quick release from isometric conditions at various moments during the twitch. It is found that the greater the amount of release during the relaxation phase, the less complete the regain of tension. The implication of this phenomenon is that during relaxation the muscle loses its waning chemical 'momentum' more quickly the more it is allowed to shorten. This is perhaps a reflection of the effect of muscle length either on the Ca-pumping ability of the triads or merely on the length of the diffusion pathway to them. Changes in either of these parameters would have marked effects on the duration of the active state, which is itself undoubtedly dependent on the free Ca level within the fibrils.

Force/velocity relations

An obvious feature of the isotonic twitches of figure 6.3 is the marked slowing down of shortening as the load on the muscle is increased. If this effect were taken to its logical conclusion and the load were increased until it just equalled the full isometric tension the muscle could develop, then, of course, there would be no external shortening at all. On the other hand, at the opposite end of the load scale, at zero load (= force), the velocity of shortening would be maximal.

The relation between load (= force) and the steady velocity of shortening is of typical exponential shape, as we see from the top curve of figure 6.5. Two forms of equation are found to fit the

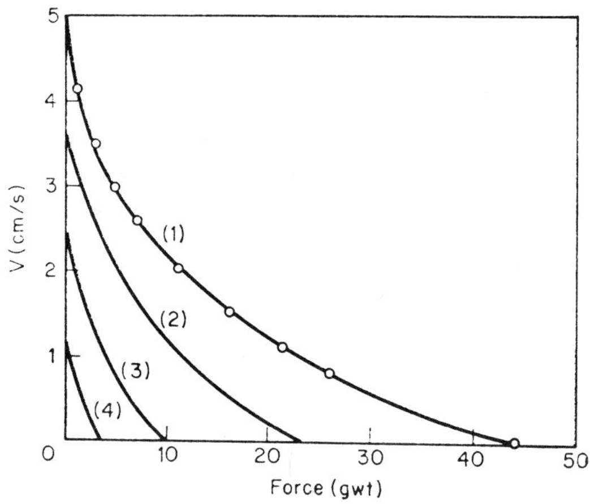

Figure 6.5. Diagram to illustrate the dependence of the velocity of shortening on the load (= force) at various stages of the isotonic twitch (so-called force/velocity curves). Frog sartorius muscle at 0°C. After [89].

Curve 1 – force/velocity curve during rising phase.
Curves 2, 3 & 4 – ditto during the relaxation phase.
Curve 2 at 460 msec – tension = 60% max.
3 at 640 msec – tension = 30% max.
4 at 830 msec – tension = 8% max.
(cf. figs. 2–4)

experimental points satisfactorily, the hyperbolic form of Hill and the exponential of Aubert, though both of them are empirical in the sense that their constants cannot easily be interpreted in terms of any real muscle parameter.

The 'characteristic' equation of Hill [67, 70, 71] is in the following hyperbolic form:

$$v = b(P_o - P)/(P + a) \qquad 6.1$$

or

$$P = (P_o + a)/(v/b + 1) - a$$

where v = velocity of shortening, P = the force (= load), P_o = maximum isometric tension the muscle can develop, b is a constant with the dimensions of a velocity and a is a constant with the dimensions of a force.

Aubert's equation [2, 3] is in exponential form:

$$P = A \exp(-v/B) - F \qquad 6.2$$

or

$$v = B \log_e [A/(P + F)]$$

where $A(=P_o + F)$ and B are constants with the same dimensions as the a and b of Hill's equation. F represents a small internal 'frictional' or compressional force, resisting shortening.

The maximal velocity, when $P = 0$, is given in Hill's equation by bP_o/a, and in Aubert's by $B \log_e A/F$. The isometric tension, P_o, when not known, is given in Aubert's equation by A–F, when $v = 0$. There is no obvious way of finding the P_o of Hill's equation, except the rather dubious one of solving 'triple' simultaneous equations from observed values of P and v. Similarly in Aubert's equation the constant F can only be found by empirical fitting of the curves.

All the force/velocity curves in figure 6.5 have been calculated by solving Aubert's equation for $F = 3$ g, and $B = 1{\cdot}7$ cm/sec, but using varying values of P_o from 44 to 3·5 g. The reason for doing this will become apparent in a moment.

The top curve in figure 6.5 is given when the isometric tension, P_o, = 44 g, and is the type of curve to be expected for a straightforward isotonic experiment in which the velocity, v, is measured from the early linear portions of the respective shortening curves, under varying loads. The circles which sit closely upon it are taken from actual experimental results in reference 89, for which the above constants were calculated by the authors. The goodness of fit evidently justifies the use of Aubert's equation.

The other curves in the figure are drawn to demonstrate what is likely to happen during relaxation from an isometric twitch, when the tension is falling, as in figures 6.3 and 6. 4, and the muscle is then suddenly released to a new and lower tension, under varying afterloads. The times at which the releases have been made are those at which the falling tension has just reached the intercept values on the abscissa ($= P_{o,t}$). Allowing for small differences in the absolute values of the velocity, these theoretical curves reproduce almost exactly an actual experiment given in reference 89, showing that Aubert's equation, and probably also Hill's, apply without any alteration of their constants, even during the decay of the active state of the muscle during relaxation.

The curves demonstrate very well what is meant by this so-called 'active state' and how it must not be confused with changes in muscle length which occur in an afterloaded isotonic twitch. For instance, the muscle in figure 6.5 is still capable of shortening at an appreciable velocity even as late as 830 msec after the shock, when the isometric tension has already dropped to less than 10 per cent of its maximal value at the height of the twitch. However, if we were to judge entirely by the isotonic experiment of figure 6.3, the muscle there seems to have relaxed completely by 600 msec even under the lightest load, so that all possibility of further shortening would have been deemed to be over. Yet there was, as we see, still appreciable activity left in it. We shall return to the concept of the active state, when discussing the tetanus.

The values for tension given in figures 6.2 onwards are those for a muscle of a particular size. To be able to compare results between muscles, it is convenient to express the tension as Pl_o/M, where l_o = rest length of muscle and M = its weight; this rules out variations due to varying cross-sectional areas. The results are then approximately in the form of tension or load (g.wt) per cm^2 of cross-section. For the frog sartorius these values can vary from 1300 to 2100 g/cm^2, partly depending on the freshness of the muscle, and partly on unknown factors, amongst which is probably the age of the animal. A good average value is 1500 g/cm^2 [3]. An equally useful but dimensionless expression is the fraction P/P_o, that is, the load expressed in terms of the isometric tension.

Elastic elements in series with contractile elements

Before describing the fusion of twitches to give a tetanus, it is important to understand the elastic properties of contracting muscle, which depend intimately on its structure. If we return to the sliding filament model (figure 1.4), we see that as soon as a contraction starts, the tension developed in the muscle substance itself must be entirely borne by the series of actin and myosin filaments and Z-discs. It is scarcely conceivable that these elements would be ideally rigid, particularly not so at the junction of the oppositely 'polarized' actin filaments in the Z-disc, and again at the points of contact of the myosin heads with actin filaments, where the active tension is developed. Hence, some 'give' will occur in these regions. Furthermore, the terminal sarcomeres in each fibril within a muscle fibre must be connected in some manner to the connective tissue of the tendons of origin and insertion, and this, too, is likely to be a relatively weak spot. Finally there will be a very slight 'give' in the tendons themselves.

Each of the components in series contributes to what is known as the series elasticity, the overall effect of the 'give' being called the series compliance. The way in which this compliance depends upon the tension is described in reference 68, but even at the very large tension of 3000 g/cm^2 obtained by stretching a muscle during a tetanus, it amounts to less than 3 per cent of the rest length. This is very similar to the compliance found in a muscle in full rigor, where many of the myosin heads will have become firmly and permanently attached to actin monomers in the actin helices [14]. It is most important not to confuse this very small compliance, characteristic of contracting muscle, with the much larger 'compliance' of resting muscle which can be stretched a long way, even under very light loads.

Besides the compliance due to the muscle and its tendons, another compliance will arise in experiments on excised muscle, due to the connections to the apparatus. This can be minimized by the use of chain, and in the most up-to-date apparatus probably does not exceed 2 per cent of the length at maximal tension. Hence with modern apparatus the total series compliance in a contracting muscle is of the order of 5 per cent, although in some earlier experiments it may well have exceeded 8 per cent.

The stretching of the series compliance is sufficiently large to have significant effects in accurate 'isometric' experiments, because even under the best conditions the contractile elements themselves will have had to shorten by this amount during the early phases, to take up the slack. This is the main cause of the difference in tension between a twitch and a tetanus. It also accounts for the very rapid initial shortening which occurs immediately after quick release from isometric conditions during a twitch [74, 89].

Fusion of twitches to give a tetanus

It is obvious from the form of the isometric twitch (figure 6.3) that further shocks delivered to the muscle at constant rate will have varying effects, depending on where they fall on the tension/time curve. At high rates of about 200 shocks per sec, delivered to the frog's sartorius at 0°C, for instance, the second shock at 5 msec will fall within the absolute refractory period of the action potential (about 10 msec in this case), and will fail to elicit any additional mechanical response. On the other hand, at the low rate of 2 shocks per second, the second shock (at 500 msec) will fall about $\frac{2}{3}$ the way through the relaxation phase, and the muscle will respond by another twitch, which will be cut off from completion by the third shock and so on. The result is a hump-backed curve (figure 6.6, curve 2), each hump corresponding to a shock [162].

At intermediate rates of stimulation between very fast and very slow, the individual twitches elicited will more and more tend to fuse. In the frog sartorius at 0°C complete fusion occurs at about 15 shocks per second, or 1 shock every 67 msec. By this stage the rising phase of the first twitch is about 70 per cent complete. The effect of fusion is to increase the active tension to 1·2 to 1·8 times the maximal twitch tension, that is between 1500 and 3000 g/cm^2 (see figure 6.6). Varying values are found for the twitch/tetanus tension ratio, the range being 0·45–0·90 [162].

The increase of tension, due to the fusion of twitches in a tetanus, arises because there is then ample time for the series elastic components to be maximally stretched by the rising tension, which is far from the case in the short-lived single twitch. It is clear that the

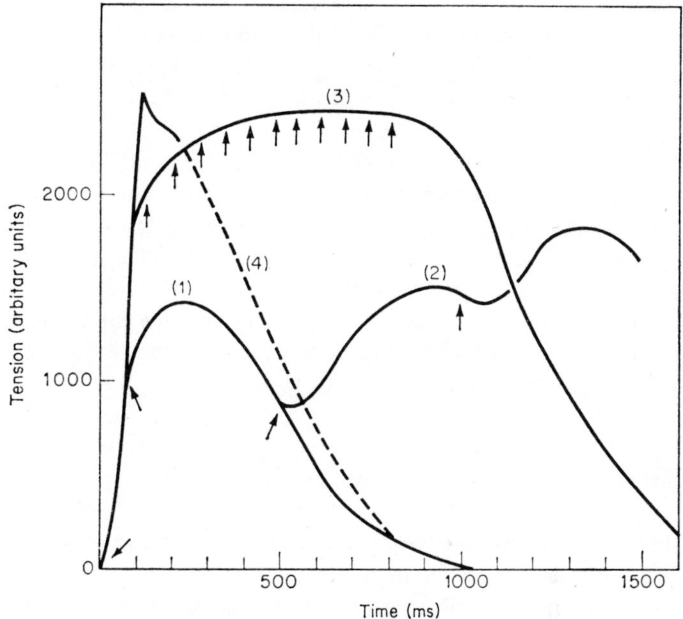

Figure 6.6. **Fusion of isometric twitches to give a tetanus.**

Curve 1 Single twitch.

2 Two shocks at 500 msec intervals.

3 1 shock every 67 msec (15 shocks/sec). Note steady build up and maintenance of tension.

4 Effect of a sudden early stretch on development of tension during a single twitch. The muscle was stretched by about 5% l_0 within 50 msec of the shock. After [68] and [162].

active state is fully established by the end of the latent period (see figure 6.2), so that if the muscle could become perfectly inelastic at that moment, the tension would rise abruptly to the maximum tetanus tension, with very little delay. In the real state of affairs, however, the rising tension first stretches the series elastic components as far as it can, thus raising their potential energy; consequently this early portion of the tension increase will not be fully registered by the external recording apparatus. The proof that this is so is provided by stretching a muscle, by about 5 per cent l_0, within the first 50 msec or so of a single twitch [68]. The tension then abruptly rises to the maximal tetanus tension (curve 4 in figure 6.6). In other words, we have done the muscle's job for it, by stretching its series elastic components, just as it reached the fully activated state.

In a sense, curve 4 can be considered a better measure of the active state during a single twitch than the simple tension curve, particularly in the rising phase. By this criterion, the active state has already begun to decline within 200 msec of the shock. The underlying chemical basis of this decline is almost certainly the withdrawal of Ca ions from the active enzyme sites on the actomyosin complex, by the action of the Ca pump in the triads of the sarcoplasmic reticulum (see chapter 4).

Effect of muscle length on tension and speed of shortening

In the experiments we have so far described, the measurements of the developed tension, or of the isotonic shortening, were made over a range of muscle lengths close to the slack length of the excised muscle or slightly longer. It is important to know what happens to these two parameters when the muscle is stimulated at much greater or lesser lengths, and whether the results bear out the predictions of the sliding filament theory.

The most elegant way of performing this type of experiment is on single living fibres, because then it is possible to measure the average sarcomere length over a marked length of fibre, by means of phase contrast microscopy, and thus to relate sarcomere length directly to the tension developed or to the rate of shortening [52]. The procedure is to dissect out a single fibre still attached to tendon at one end and bone at the other, and then to mount it in a horizontal perspex bath. The bone end is held rigidly, and the tendon end connected to a miniature tension transducer. During the preparation, the fibre has placed upon it, and fixed with grease, two miniature gold foil markers, towards its centre. These markers are used to reflect a beam of light back through a microscope lens, and thence to a cathode-ray-tube spot-follower in one direction at right angles to the fibre axis, and to a pair of photocells in the other. The impulses from the spot-follower and the photocells are fed through a complex amplifier circuit, and then to a small servo-motor, attached to the bone end of the fibre. It is then possible by means of impulses to the servo-motor to keep the distance between the two gold-leaf markers on the fibre constant at a pre-fixed value during an isometric tension experiment, or to use the device to follow length changes, during isotonic experiments [52].

In most of the experiments to be described, the fibre was placed in oxygenated Ringer's fluid at 0 to 3°C, and stimulated at 20 shocks per second, to give an isometric tetanus, with the sarcomere length fixed at pre-determined values. The results, plotted as relative isometric tension against sarcomere length, are shown by the solid line in figure 6.7*a*, for a fibre from a frog semitendinosus muscle. In this muscle, the actin filament lengths from tip to tip across the Z-disc, and the myosin filament lengths, are accurately known (2·05 and 1·60 μ, respectively), and so for a given sarcomere length it is possible to show the positions of these filaments relative to one another, as in figure 6.7*b*.

By cross reference from figure 6.7*a* to figure 6.7*b*, we see that there are five critical sarcomere lengths on the length/tension curve: A, very long (3·65 μ) where the developed tension is zero; B and C, near to the slack length (l_o), where the tension is maximal; D, a point at which overlap of actin filaments from each half sarcomere has begun; E, the point at which the Z-discs from each half sarcomere have just been pulled into contact with the ends of the myosin filament; and F, another point of zero tension, at which the myosin filaments have become buckled by pressure from the Z-discs, through which they cannot pass.

Thus the inflections in the length/tension diagram are evidently directly related to critical points in the sliding of the filaments over one another. At 3·65 μ, for instance, the sarcomere is so long that

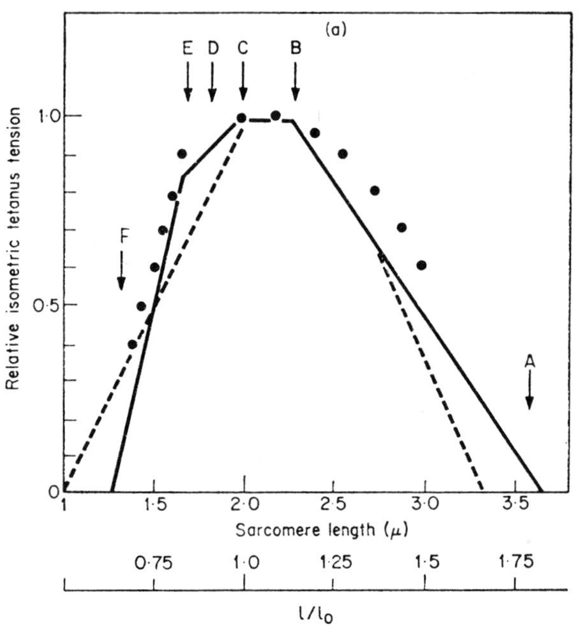

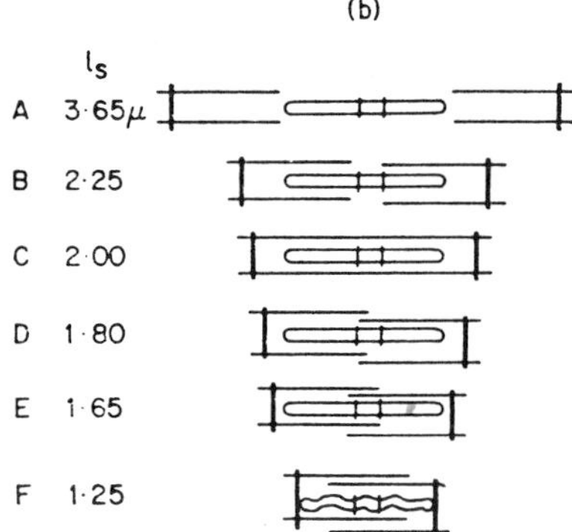

Figure 6.7. (*a*) Composite diagram to show effect of sarcomere length on the tension developed by single muscle fibres (full line and dots), after [42] and [52]; also superimposed as dotted curve, results for an intact sartorius muscle [3].

(*b*) Diagram to show overlap of myosin (thick) and actin (thin) filaments at various sarcomere lengths, for comparison with companion figure. After [52].

the actin filaments are scarcely in contact with myosin, and hence no tension can be developed. As the sarcomere is allowed to shorten from this length down to 2·25 μ, the tension increases in proportion to the sarcomere length, because increasing numbers of myosin heads are able to get into contact with actin and so develop active tension. Further shortening to 2·0 μ is without effect, because now the tips of the actin filaments are moving over the central bare patch of the myosin filaments, where there are no heads (cf. figure 6.7*b*). From this point downwards to 1·65 μ, the tension slowly declines to about 0·85 of maximal, presumably because the actin filaments from each half sarcomere have increasingly overlapped one another, and therefore interfere in the formation of myosin-actin links. At 1·65 μ, the critical point is reached where the Z-discs have just banged into the ends of the myosin filaments, and from this point onwards the tension declines dramatically, as compression and buckling of the myosin filaments increases, until at 1·25 μ it reaches zero. From these elegant experiments alone, we have the strongest possible confirmation of the sliding filament theory, with the added safeguard absent from much earlier work, that the sarcomere length was accurately controlled during the tetanus.

At about the time the above results were published, a similar set of experiments on frog semitendinosus fibres, from another laboratory, appeared in the same journal [42]. The dots in figure 6.7*a* are taken from the latter paper, and it is seen that they agree closely, the small differences between the two probably being due to rather less accurate control of sarcomere length in the latter experiments. Also included in the figure (dotted curve), are some results on the isometric tension developed during a tetanus of a whole frog sartorius muscle [3]. Here, the sarcomere length was not measured, but by putting the slack muscle length (l_o) equal to a sarcomere length of 2·0 μ, the curve agrees remarkably well with those for single fibres.

Another outcome of the first set of single fibre experiments was the clear-cut demonstration that the velocity of shortening in lightly-loaded *isotonic* twitches was virtually independent of sarcomere length from initial lengths of 2·6 down to 1·65 μ. Below this point, increasing buckling of the myosin filaments occurs by pressure from the Z-discs, as we have seen, and this results in a sharp decline in the velocity of shortening. The development of this new force, resisting shortening below 1·65 μ, means that then neither the measured isometric tension, $P_{o,l}$, nor the nominal load during isotonic experiments, represent the true force the contractile elements themselves develop in this shortened state. Under such conditions, a new force must appear in both Hill's and Aubert's equations, although it is not known at present whether it is constant or length-dependent [52].

Work done during a twitch or a tetanus

From the preceding sections it is clear that the work done during a single twitch ($= W = -P\Delta l$) will depend on the load, P, in a far from linear manner, because both the duration and the velocity of shortening (v) are themselves dependent on the load. Employing Hill's equation, 6.1, for the dependence of velocity on load we see that during the early phase of the twitch, when v is virtually constant (cf. figure 6.3), the shortening, during time t, $= vt$ and the work done $= Pvt$, and hence:

$$W = Pvt = bPt(P_o - P)/(P + a) \qquad 6.3$$

This equation gives a typical bell-shaped curve for the plot of W against P with two zero points for W when $P = 0$ and when $P = P_o$, and a maximum when $P = \sqrt{a(P_o + a)} - a$. Since the 'constant', a, is found to vary between 0·25 to 0·4 P_o, it follows that W is maximal when $P = 0·31$ to $0·35\, P_o$ or when $v = 0·31$ to $0·35\, v_{max}$. A plot of the work done in unit time under varying load, as predicted by this equation, is given in figure 6.8 (full curve).

The above equation cannot be used for estimating the total work done during a twitch, because it applies at best only during the early linear phase of shortening, particularly under the heavier loads. In such heavily loaded twitches, the velocity of shortening remains constant only for a limited time, and then falls more or less abruptly to zero, as we see from the curve for the 9 g load ($P/P_o = 0·75$) in figure 6.3, where the velocity was constant until only 200 msec. On the other hand, in the most lightly loaded of the three twitches in the figure (3 g load = 0·25 P/P_o), the velocity remained constant for a further 100 msec, that is until 300 msec. The work done is thus unduly curtailed under heavy loads, and this is exacerbated by the fact that the onset of external shortening also becomes more and

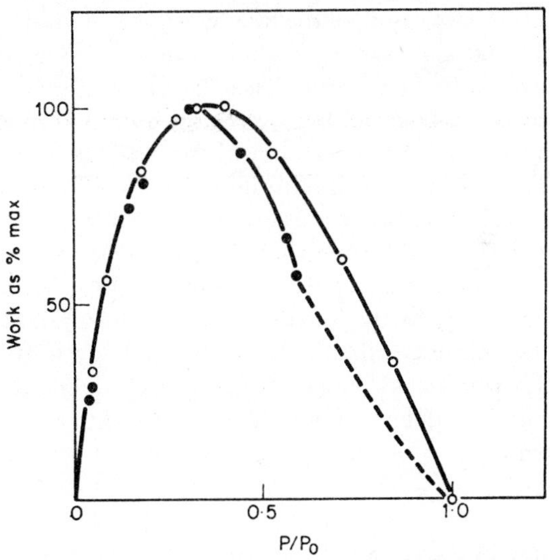

Figure 6.8. **To show dependence of the work done, during a single twitch, on the load on the muscle (P/P_0). Frog and toad twitches at 0°C shown as dots. After [71]. The circles and the full line through them show the theoretical work done in unit time according to Hill's equation 6.1, with $a/P_0 = 0{\cdot}3$.**

more delayed, as the load is increased, because time is required to stretch the series elastic components first. The overall effect of these factors is best illustrated by plotting the actual total work done against the relative load, and comparing it with the theoretical work which would be done in unit time at steady velocity according to Hill's equation (figure 6.8). There is good agreement between the values at loads below $P/P_o = 0{\cdot}40$, whereas at higher loads there is an increasing discrepancy, because of the factors we mentioned.

During a twitch it is often difficult to obtain an exact measure of the internal work done by the muscle in stretching its own series elastic components, which becomes an increasingly important factor the greater the load (cf. [24]). Sometimes an approximate value for this can be obtained, if the compliance of the muscle and its connections is exactly known, and there are also roundabout ways of calculating it from the delay in onset of shortening [69]. For this reason, both Hill's and Aubert's equations in their simple form fail to give a correct measure of the total work at high loads.

The difficulties of assessing the total external and internal work done, particularly under heavy loads, largely vanish when an isotonic tetanus is substituted for the twitch, because then time is no longer of such importance, and the early phases dwindle into insignificance. For example, a frog sartorius under a load of 0·6 P_0 can do about 40 g.cm of work per g in a single twitch, whereas in a tetanus of 1·3 sec duration it does nearly 300 g.cm. In fact, the work done during a tetanus is limited only by the muscle's inability to shorten much below 0·6 l_0, for obvious reasons (see figure 6.7*b*). The tetanus also has the advantage that more accurate values can be obtained for the maximal rate of doing work (= power), because the velocity stays constant for longer. Both Hill's and Aubert's equations then apply accurately.

Is relaxation an active process?

On the basis of the sliding filament hypothesis, the relaxation process would not be expected to be an active one, in the sense of active pushing by the micro-filaments. In experiments with isolated model systems of fibres and fibre bundles, for example, some force has always to be applied to stretch the muscle out under relaxing conditions (≡ inhibition of ATP-ase activity). Yet with intact muscles, particularly those of mammals such as the sterno-mandibularis of the ox, it often happens that the muscle relaxes, apparently spontaneously, from a series of twitches or from a tetanus, even when it is laid out horizontally on a plate. It is observations of this sort which have given rise in the past to the persistent belief that relaxation is an active process, in the sense we described.

Apart from the results with model systems, there are two other lines of evidence against the idea that relaxation is 'active'. First, muscles which show spontaneous relaxation to a marked degree are always those with a very high connective tissue content, such as the sterno-mandibularis we mentioned above, which contains 10 per cent collagen on the dry weight basis. On the other hand, the sartorius of the frog, which has much less connective tissue, can be contracted down to very short lengths by a series of twitches and will usually only lengthen again when it is pulled out forcibly, albeit the force required is small [69, 71]. Even this muscle tends, however, to be pushed out again by its connective tissue, if it is under Ringer's fluid. This suggests that the 'force' involved when spontaneous relaxation occurs arises entirely from extra-fibrillar connective tissue, coiled up into spring-like structures by active shortening of the muscle, and uncoiling again spontaneously when the active tension developed by the contractile filaments decays away. The other line of evidence against active participation by the contractile filaments comes from studies of living single fibres, where it is possible to strip off the sarcolemma and then cause the fibre to contract by injecting a minute amount of Ca ions [124]. Such fibres refuse to relax spontaneously, although they can be easily pulled back to their original length, once the sarcoplasmic reticulum has had time to pump away all the injected Ca^{++}.

Resting tension

Apart from the active tension a muscle can develop when it is stimulated, its tension in the resting state (P_r), rises progressively as it is stretched beyond l_o, according to an exponential equation of the form:

$$P_r = ke^{l/cl_o} \qquad 6.4$$

or

$$l/l_o = c\,(\log_e P_r - \log_e k)$$

Because of the size of the constants, c and k, in this equation, it is found that the resting tension increases from zero at l_o to only about 3% of the full active isometric tension, at $l = 1{\cdot}25\, l_o$ [69]. Beyond this length, resting tension increases rapidly, and soon ceases to obey the equation. Indeed, under extreme conditions, at a length of about 1·65 l_o, when the actin filaments have been pulled nearly out of contact with myosin, the resting tension rises so far that it exceeds the active tension the muscle can develop at l_o, that is, it becomes greater than 1500 g/cm² or so. Typical load/extension curves are

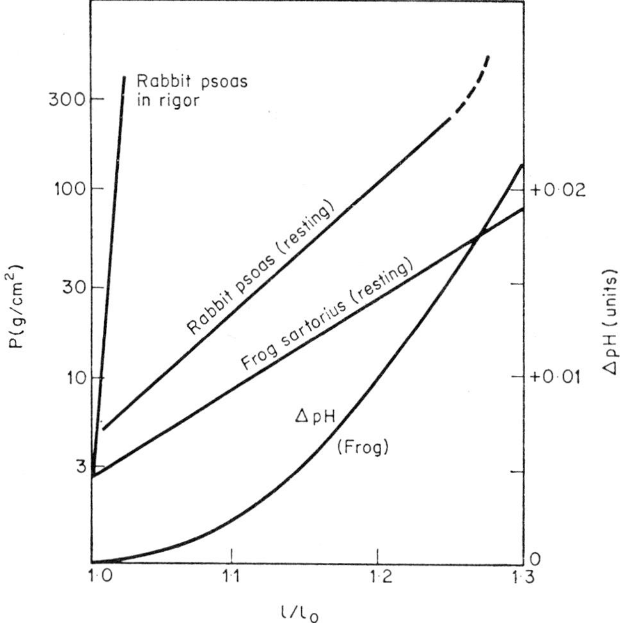

Figure 6.9. **The dependence of resting tension of a muscle on its relative length (l/l_0). Frog sartorius and rabbit psoas muscles are shown, the latter in the resting and rigor states ([69] and author's observations). Also shown are the pH changes when a frog sartorius muscle is stretched (after [32]).**

given for frog and rabbit muscle in figure 6.9. The increasing resistance of the resting muscle to extension at great lengths is due to elastic elements in parallel with the contractile filaments; the latter themselves offer almost no resistance in the resting state [69, 124], as we have seen.

The parallel elastic elements are of two sorts: those within the fibre, mainly the sarcoplasmic reticular vesicles; and those in the fine connective tissue fibres of the sarcolemma, and the thicker connective tissue surrounding the bundles of fibres (peri- and epimysium; see figure A.ii). By careful dissection of the sarcolemma from living fibres, it can be shown that very little of the resting tension can be attributed to this structure, or even to the coarser connective tissue, in the length range from l_o to 1·30 l_o. Within this length range, it seems that the tension must chiefly be borne by the tubules of the sarcoplasmic reticulum, unless of course we care to believe that there really is a highly elastic protein, such as the postulated myofibrillin [54], connecting the ends of the actin filaments from each half sarcomere. Such a linking protein would have to possess extremely low density, rather like the resilin of insect flight muscles, in order to have escaped detection by electron microscopy for so long [170].

Beyond about 1·3 l_o, it is not difficult to account for the increasing resting tension which is undoubtedly borne mainly by the sarcolemma, and also by the collagenous connective tissue of the peri- and epi-mysium. This is substantiated by the heat changes which occur during stretching: up to 1·30 l_o, muscle possesses rubber-like elasticity, so that it gives out heat, as it is lengthened, and takes in heat again, as it is allowed to shorten passively back. This is due to a decrease in configurational entropy (ΔS) on stretching, which leads to a heat output equivalent to $-T\Delta S$; that is, the molecules undergoing the stretch become increasingly ordered and this process is reversed on release. Beyond 1·30 l_o, the heat output changes sign until at about 1·40 l_o, the muscle takes in heat as it is stretched further. This is akin to the small thermoelastic effect [165] found on stretching a contracting muscle, and probably arises from attempting to stretch the taut helical structure of proteins such as collagen.

There are two other important characteristics of the stress/strain curves of resting muscle: (i) there is a time factor involved, in the sense that an extra load on the muscle causes a rapid initial extension followed by a slower creep, or if tension is being measured, by a rapid initial rise in tension followed by a fall back to a steady value; (ii) there is a more or less prominent hysteresis effect, in the sense that the length/tension curve for increasing lengths gives greater tensions than the return curve for decreasing lengths. Both of these features might be expected in the early part of the stress/strain curve, where it is probable that the irregular longitudinal tubules and vesicles of the sarcoplasmic reticulum are the main elements deformed by the stress. The viscosity of the sarcoplasmic fluid through which the actin and myosin filaments must slide backwards and forwards during the deformation must also contribute to this so-called 'damped' elasticity.

Another interesting phenomenon is the positive pH change which accompanies stretching of a resting muscle [32]. This can be measured by surface glass electrodes. The size of the change is indicated in figure 6.9, where it is plotted against length, for frog sartorius muscle. There is no obvious explanation of it at present, although it may arise from stretching of the sarcoplasmic reticulum, which causes internal osmotic pressure changes and thus a movement of water and ions.

7: Energetics of Contraction

The subject of energetics covers the whole spectrum of energy changes in muscle, from the arrival of the action potential at the triad junctions, through the release of Ca ions into the sarcoplasm, to the onset of the breakdown of ATP at the active sites on the actin and myosin filaments, and the subsequent transduction of the chemical energy arising from this source into work or the development of tension. To simplify it, we shall discuss only the beginning and end result of the process here, that is to say, the initial activation via calcium release, and the later performance of work, development of tension and output of heat; we shall come back to the difficult and disputed area of energy transduction in the next chapter. Our main aim is therefore to relate the observed output of heat to the work done or the tension developed.

When a muscle contracts isotonically and does work in lifting a load, it gives out simultaneously a certain quantity of heat. Heat is also given out under isometric conditions, when the only work the contractile filaments are doing is the small amount needed to stretch the elastic elements in series with them, including the connections to the recording lever. Even after these elastic elements have been fully stretched in a sustained isometric contraction, there is still a continuous to-and-fro movement of the actin filaments in relation to the active myosin heads, which has been described vividly as a kind of dithering [31]. During this dithering, the small amount of work done during the shortening phase of each miniature cycle is reversed again during the lengthening phase and, therefore, does not appear in the overall energy balance sheet.

During the later process of relaxation from an *isotonic* contraction, the main physical change is the falling of the load and lengthening of the muscle, as the active state declines, due to the withdrawal of Ca ions; in an *isometric* contraction, on the other hand, most of the lengthening during relaxation occurs internally, as the stretched elastic elements shorten and pull out the relaxing contractile filaments once again. In both cases, the work changes sign during relaxation: work is now done *on* the muscle by the falling load or the shortening elastic elements. All of this 'negative' work becomes degraded into heat, of which a small part comes from the load hitting the afterstop and the remainder from the muscle itself. This latter, major part of the heat is captured by the thermopiles used to measure the heat changes.

The first law of thermodynamics

All these energy fluxes come within the compass of the first law of thermodynamics, which states that the total energy of a system and its surroundings must remain constant: an alternative form of the law is that, *whenever a quantity of one kind of energy disappears an exactly equivalent amount of other kinds must be produced* [51, 163]. The first law can be expressed very simply for our purpose, because muscle is a system where the volume remains sensibly constant even during activity, so that its internal energy is essentially the same as its enthalpy, or heat content. We shall denote this internal energy as E, to distinguish it clearly from the heat which is given out; many treatises prefer the symbol H, but this is merely confusing in this context. It then follows:

$$-\Delta E = W - Q \qquad 7.1$$

where W is the work done by the muscle, and $-Q$ is the heat given out.

The first law, in this form, expresses the fact that the muscle has done an amount of work, W, during the performance of which it has given out an amount of heat, $-Q$, and lost internal energy proportionately. Note that we can only describe *changes* in the internal energy and not its absolute quantity which is unknown, but very large. Another term can be defined at this stage, the so-called *mechanical efficiency*, which is given by $-W/\Delta E$.

Because of the nature of the muscle twitch and tetanus, it is possible with very refined heat recording apparatus to divide the heat term, $-Q$, up into a number of others [71, 72]. Thus, it is possible to distinguish very rapidly appearing early heat which is given out before there are any signs of the development of tension or of external shortening [73, 74]; this heat probably arises from the activation process, that is, the release of Ca ions from the triads and their subsequent capture by the active sites on actin and myosin. It will be called here *activation heat*, defined as $-A'$. Then, as the muscle begins to contract and do work, further heat is given out because the transduction of chemical into mechanical energy is not ideally efficient. This heat we shall call contraction heat, $-Q_c$. Last of all, heat will be given out during relaxation, mainly from the degradation of work, but there is also the possibility that reversal of the chemical events of contraction gives rise to small heat changes. The latter 'chemical' part of the relaxation heat, not easily distinguishable, we shall call $\pm Q_r$; its sign may be positive or negative. The overall energy flux during a contraction can then be written:

$$-\Delta E = -A' - Q_c + W \pm Q_r \qquad 7.2$$

Note that the sign of Q_c is always negative, i.e. heat is always given out in the contractile phase. As we noted, the sign of Q_r is unfortunately still not known.

The outcome of equation 7.2 is as follows: if the heat changes are measured during the contractile phase of the twitch or tetanus only, then the heat output $= -A' - Q_c$ and the total energy flux appears to consist of $-A' - Q_c + W$; on the other hand, if the heat is measured up to the end of relaxation under a *falling load*, then the total output of heat is approximately equivalent to the total energy flux, that is, to all the terms on the right of the equation. This is because all the work done has been reversed and thus reappears as heat, when the load falls; most of this is measured by the recording apparatus. Another way of studying relaxation is to hold the load up at the end of the rising contractile phase, so that it cannot pull out the muscle again. In such a case, the work done is not degraded into heat, so that the heat changes often appear to cease entirely at the end of the rising phase, although there is sometimes a small absorption of heat at this time. A little thought will show that equation 7.2 is still approximately obeyed under these conditions too [24, 71, 163].

It is most important that the reader understand a fundamental aspect of a twitch or a short tetanus, before proceeding further. This is that the primary process of ATP breakdown, and its more or less immediate resynthesis from creatine phosphate (PC), are the dominant energy-yielding chemical changes of the contractile and relaxation phases; the resynthesis of ATP and PC from the glycolytic cycle under anaerobic conditions, or by oxidative rephosphorylation under aerobic conditions, are very much delayed processes, the heat from which appears, in varying degree, several hundreds of milliseconds after the completion of relaxation. The only immediate need of the contractile filaments in the relaxation phase is a supply of ATP to act as a plasticizer.

Measurement of heat changes

The measurement of heat changes has for many years been refined to a point which is unlikely to be much improved upon, because of the restrictions imposed by the material itself. It consists in the use of a micro-thermopile, usually placed between a pair of muscles from the same animal, as shown in figure 7.1 [70, 71]. The thermopile itself is about 12 μ thick, protected by a total of 33 μ of mica and bakelite (I. *a* and *b*), and is made of about 40 constantin-chromel couples. It fits into brass flanges in the muscle chamber as shown in II, the complete set-up in the muscle-bath being shown in III. The pair of muscles fit on either side of the groove, *c*, in II, and are connected to the chain and level system. The stimulating electrodes, *m*, are placed at either extremity of the muscle chamber. The most important features of the thermopile are the three insulated dummy couples at the fixed end of the muscle, (*t* in IV) and the twenty dummy couples at the end attached to the chain (*v* in IV). This means that when shortening occurs only muscle which has been under the same thermal conditions as those at the centre can be pulled from the direction of *h* (in III) into the region of the measuring couples. If such dummies were not included, it would mean hot muscle, coming from the region outside the chamber, would pass into contact with the active couples, and produce an artefact.

The current changes produced in the thermocouples are recorded

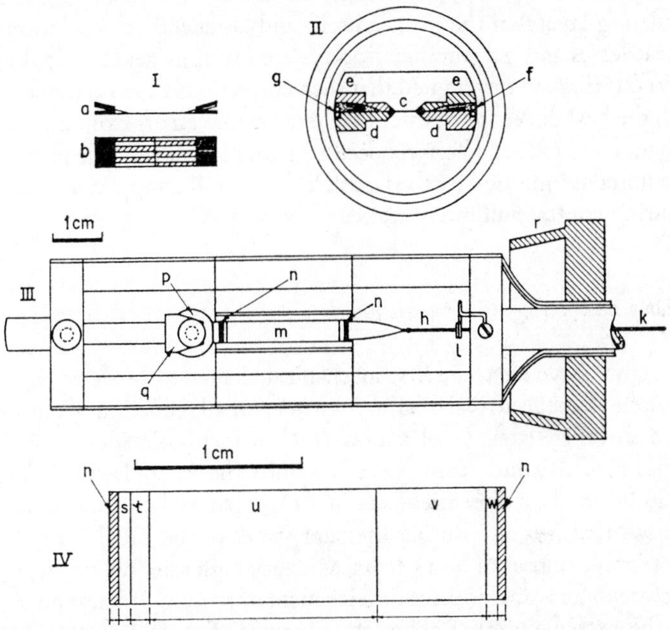

Figure 7.1. Thermopile for measuring muscle heat. I(*a*) end view and I(*b*) side view of element unmounted. II, end view of element mounted: hot junctions in middle of groove. III, complete instrument with pair of muscles: *m*, clamped at *p* and connected by thread and chain (*h*, *k*) to recording lever: *n*, stimulating electrodes. IV, enlarged view of portion of element in groove: *u*, 42 insulated couples, protected by *v*, 20, and *t*, 3 insulated dummies. After [70] and [71].

on a specially designed galvanometer, and finally displayed on a cathode-ray oscillograph. A further refinement is to build an automatic differentiating system into the amplifying circuit, so that the heat changes are displayed as $-\mathrm{d}Q/\mathrm{d}t$, instead of the integral, $-Q$. Further details of the apparatus are given in [3] and [70, 71]. The normal method of treating the muscles is to place them in oxygenated Ringer's fluid for some time before an experiment begins. The bulk of the Ringer is then removed, leaving an adhering surface film on the muscles, when they are in position on the thermopile.

Most of the heat measurements to be discussed were made with pairs of frog sartorius muscles, weighing about 0·08 g, about 3·2 cm long, and about 0·035 cm thick. Complications arising from insulating layers on the muscle itself, and a general consideration of heat losses and galvanometer delays are given in detail in [70] and [71]. It is taken for granted that such corrections have been made to all the heat curves to be discussed here, so that errors from measurement artefacts have been reduced to a very low level. The precision required is indicated by the fact that the overall temperature change during contraction rarely exceeds 3×10^{-3} °C.

Early heat during an isometric twitch

As we have seen, the first mechanical change to be detected in a muscle after the arrival of an AP is a very small reduction of tension, the so-called latency relaxation (LR), which coincides with the release of calcium from the triads into the sarcoplasm. A large number of heat measurements in frog, toad and tortoise muscles shows that it is also during the later stages of the LR that the first detectable output of heat occurs, at a very high rate [67, 71, 72, 73], before there is any positive development of tension. It is not possible to be extremely exact about the moment of onset, because of the difficulty of correcting accurately for galvanometer lag and lag due to the thin inert layer on the outside of the muscle, during the first 10 to 20 milliseconds of the twitch. The general pattern is shown by the results given in table 7, for a number of different muscles: the first heat change is detectable one third to half way through the mechanical latent period before any positive tension has been developed, and rapidly attains its maximum rate. Even before positive tension development can be detected, the rate has already begun to decline, and has fallen to less than half maximal at about the time half the maximal tension has been developed.

It is possible with toad sartorii to compare the heat changes with the liberation of Ca ions into the sarcoplasm, from the results of two investigators [72, 90]. After due allowance for temperature differences between the experiments, the results shown in figure 7.2

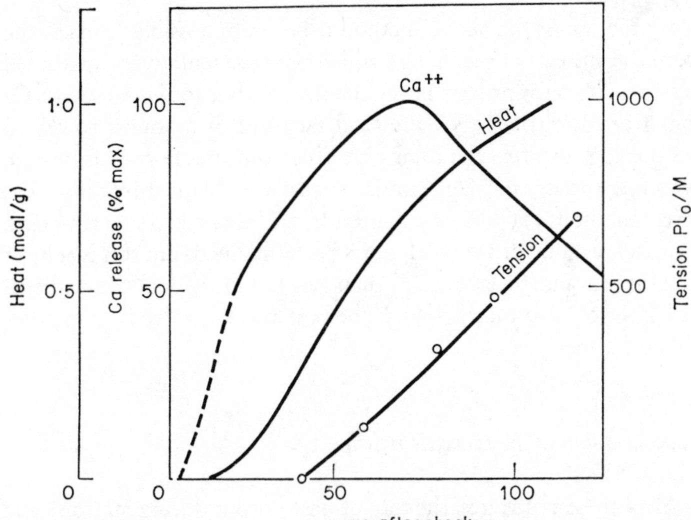

Figure **7.2. Composite diagram to show the release of Ca ions, the heat output and the development of tension during the very early stages of a twitch in a toad sartorius muscle at 0°C. Max. tension = 1200 g/cm²; heat at height of twitch = 3·0 mcal/g. See [72], [73] and [90].**

are obtained; from this we see that Ca release seems to precede even the heat change, although the fastest phase of the latter occurs well before the maximum of the Ca curve has been reached. It is not wise to draw too far-reaching conclusions from the figure, because of uncertainties in the first few milliseconds after the shock, but it is at least plausible that the early rapid phase of heat output arises directly from the process of Ca release. We shall return to this question when we come to consider the chemical events in more detail.

Table 7 illustrates another interesting point about the early heat: its onset can be delayed and its rate lowered by increasing the strength of the bathing Ringer solution. This treatment has an even more drastic effect on tension development, not only delaying its onset as shown, but also its extent. In fact, on increasing the strength of the Ringer to 2·8 × normal, a toad's sartorius will develop no tension, but the early heat is still evolved, apparently quite normally [73, 74].

By the use of the above method it becomes possible to assess the extent of the early heat change which we have tentatively attributed not to a chemical process in the strict sense, but to the release of Ca ions from the triads. For the toad sartorius, it amounts to 0·6 to 0·8 mcal/g, whereas the total heat given out in a normal isometric twitch is about 3 mcal/g. Similar results are obtained by rather less direct methods [71]. More recently, by the use of two superimposed twitches, a value of 1·2 mcal/g has been obtained, but this is subject to a large standard error, and many of the individual results lie in the range 0·8–0·9 mcal/g [50]. The best average value is, therefore, about 1·0 mcal/g.

Later heat during the isometric twitch

After the early stages, the rate of heat output during an isometric twitch falls rapidly, as we see from table 7 and figure 7.2. The full extent of the changes during a complete twitch of a toad semimembranosus muscle is shown in figures 7.3*a* and *b*, where the rate of heat output from 100 millisecs onwards is given for comparison with the tension change. In this case, the differential curve ($-\mathrm{d}Q/\mathrm{d}t$) was calculated over small time intervals from an integral curve, published elsewhere [71].

The significant features of figure 7.3 are first, the very rapid fall of the heat rate, from an initial value of about 19 mcal/g/sec, not shown on graph, to about 1 mcal/sec by the time the tension has reached its peak (1000 msec after the shock). This phase is followed by a plateau lasting 300 msec, during which relaxation begins. By the time relaxation is in full swing at 1400 msec, however, the heat rate suddenly jumps up to a new peak, and then gradually falls to zero, as relaxation is completed. The jump in heat rate is well brought out by plotting $-\mathrm{d}Q/\mathrm{d}t$ against tension in the relaxation

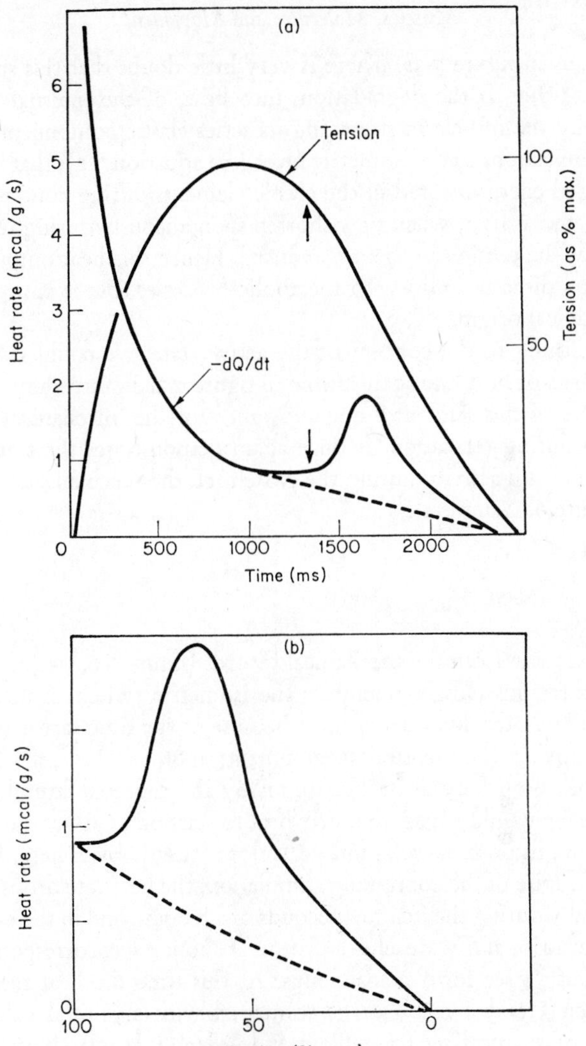

Figure 7.3. (a) The rate of heat production ($-dQ/dt$) during an isometric twitch of a toad semi-membranosus muscle at 0°C. Tension development shown for comparison. After [71].

(*b*) Plot of the heat rate against the tension during relaxation.

Dotted line in both figures represents the likely heat rate in an ideal isometric twitch, where there would be no stretching of the series elastic components.

phase, as in figure 7.3*b*. There is very little doubt that this sudden jump is due to the degradation, into heat, of the internal work done by the muscle in stretching its series elastic components and the connections to the isometric lever (see equation 7.2); that is, the potential energy, stored in the elastic elements during contraction, is released as heat, when they shorten spontaneously during relaxation, as the tension in them is reduced. Hence this heat output has nothing directly to do with metabolic processes, but is entirely of mechanical origin.

Returning to the concept of the active state, we could say that the curve of heat rate against time in figure 7.3*a* is a tolerably good measure of this state after due allowance for the 'mechanical' heat bump during relaxation. It once again demonstrates the transient nature of full activity during the twitch (cf. the mechanical records in figure 6.6, curve 4).

Isometric tetanus

As we saw from the mechanical records (figure 6.6, etc.) the isometric tetanus closely resembles the isometric twitch in its early stages, but later divagates from it because of the time factor which then allows greater tension development; in other words, the active state has been kept at its maximum by the repeated stimulation. Hence we would expect the early heat to start off at about the same high rate in both cases, as indeed it does [3, 67]. Even here, however, in spite of the continuing stimulation, the heat rate drops very markedly during the first few seconds of a tetanus, and in the case of frog sartorius at 0°C, reaches a plateau at about 7 sec, corresponding to about 14 sec for a toad tetanus. At this time the heat rate lies between 3 and 4 mcal/sec/g, compared with an initial value in excess of 35 mcal/sec (cf. table 7 and [3] and [165]). Note that, for comparison with the toad twitch in figure 7.3*a*, the steady rate in a toad tetanus is 1·5 to 2 mcal/sec/g.

The fall in heat rate can be expressed in the following exponential manner by Aubert's heat equation [3]:

$$-\mathrm{d}Q/\mathrm{d}t = h_A \exp(-at) + h_B \qquad 7{\cdot}3$$

where h_A = 'labile' heat rate, h_B = steady heat rate and α = a constant.

Typical values of the constants for frog sartorius at 0°C are 5·3 and 3·9 mcal/sec/g for h_A and h_B respectively, giving a calculated initial rate of 9·2 mcal/sec/g ($= h_A + h_B$), which is only $\frac{1}{4}$ of the true initial value. The constant α is of the order of 0·74 sec^{-1} ($1/\alpha = 1{\cdot}35$ sec).

Aubert's equation, although empirical in the sense that the initial heat rate given by it is too low, serves a useful purpose in describing the heat evolution during a long tetanus, and makes possible the assessment of the effect of initial muscle length on this evolution.

Effect of muscle length on heat output

We have seen how the initial length of a muscle has marked effects on the tension it can develop (figure 6.7); maximum tension is developed at a sarcomere length of about 2·2 μ, where the actin filaments from each half sarcomere are nearly touching in the centre of the A-band and are therefore in contact with the maximum possible number of myosin heads. At greater lengths the tension falls, because less and less actin will be in contact with myosin, until at about 3·65 μ, it has been pulled completely out of contact, and the tension falls to zero. On this side of the length/tension diagram, the steady heat rate (h_B) falls off in a very similar manner to the tension, whereas the 'labile' rate (h_A) is much less reduced, as we see from figure 7.4 [3].

Below l_0, both h_A and h_B fall off more slowly than tension as the length is reduced [3]; it is here that more and more overlap is occurring between actin filaments from opposite halves of the sarcomere, until the A-band finally begins to be compressed by the Z-discs [52]. On this side of the length/tension diagram the ratio h_A/h_B is nearly constant, irrespective of length. The final result is that the heat rate is still above 50 per cent of the maximum at lengths so short that tension has been reduced to zero.

The exact meaning of the effect of initial length on the heat rate is still not clear, though it appears likely that the steady rate of heat liberation, h_B, is concerned with the maintenance of tension; that is, it arises from interactions between the actin and myosin filaments, since it falls to zero when these filaments are pulled out of contact with each other at great lengths. At short lengths, on the contrary,

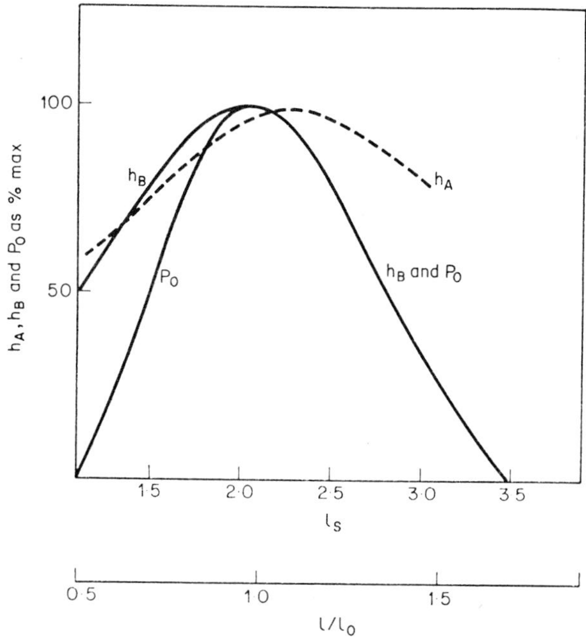

Figure 7.4. Effect of relative muscle length (l/l_0) on tension (P_0) and heat rate (h_A and h_B) during a long tetanus of a frog sartorius muscle at 0°C. Values from table XVI, in [3].

h_B is not as much affected by length as the tension is, probably because interactions between actin and myosin can still occur at these very short lengths and give rise to heat, whereas tension cannot be developed because of the mechanical resistance of the myosin filaments to compression by the Z-discs. In this respect, we can draw an analogy with the splitting of ATP by myofibril suspensions, discussed in chapter 2. Fibrils contract to very short lengths within a second or two of the addition of ATP, but they are nevertheless capable of splitting ATP at a high rate for a long time after this contraction is complete. The analogy is made more realistic by the fact that the splitting of ATP by frog myofibrils during the first 10 sec at 0°C could give rise to about 1·7 mcal of heat/sec/g of muscle, assuming 10 kcal of energy are liberated by the hydrolysis of 1 mole of ATP (cf. table 2). A good average value for the liberation of heat in a 10 sec isometric tetanus of living muscle at very short lengths is 1·8 mcal/sec/g [3].

An alternative hypothesis to account for the steady heat rate (h_B) during a tetanus is that it arises from the sporadic liberation of Ca ions from the triads, together with the opposing process of actively pumping these ions back. If this is true, it is difficult to see why it should be so drastically affected by muscle length, particularly beyond $l/l_o = 1$ [134]. On the other hand, the pumping hypothesis could partly account for the effect of length on the 'labile' rate, h_A, which is by no means as dramatic. We could suppose that this rate, which is a reflection but not an exact measure of the true initial rate of heat liberation, represents the after-effects of the activation process, which involves Ca liberation from the triads. It is unwise, however, to take the argument further in the absence of critical experiments.

Isotonic tetani and the heat of shortening

In the isometric experiments we have just described, shortening of the contractile elements was restricted to the small amount, not exceeding 7 per cent l_o, needed to stretch out the series elastic elements. Any heat specifically associated with this shortening would not have been detected amidst the large energy flux from other sources. In order to demonstrate whether extra heat is indeed released when a muscle shortens, it is therefore necessary to have recourse to isotonic experiments where shortening can be controlled accurately.

The simplest type of isotonic experiment is one in which a muscle is tetanized and allowed to contract isometrically for a short time, before being suddenly released to a pre-determined shorter length [3, 67, 74]. When this is done, it is always found that extra heat is given out over and above the 'isometric' heat, and that the time course of this extra heat liberation closely follows that of the shortening. A typical experiment is illustrated in figure 7.5. Here, the muscle was tetanized and held isometrically for 400 msec, before being suddenly released from 1·1 l_o to 0·9 l_o, a total distance of 0·2 l_o. In experiment 1, a light load of 0·05 P_o (=100 g/cm²) was used, whereas in experiment 2 the muscle shortened against a ten times heavier load (1000 g/cm²). Although shortening was much faster in experiment 1 than in experiment 2, the extra heat given out over and above the isometric amount, accompanied the shortening in both cases, and pursued a similar time course to it. However, as soon as the shortening came to its pre-determined end, the slope of the heat output curves changed back abruptly to the lesser slope characteristic of isometric conditions; in other words, extra heat production ceased as soon as shortening ceased.

A surprising feature of the curves in figure 7.5 is that more 'extra heat' was given out in the more heavily loaded contraction, 2, than in the lighter, 1, despite the fact that the amount of shortening was identical in both cases. This effect can be expressed in approximate mathematical form [67, 74], in terms of the load on the muscle:

$$-Q_s = -(k_1 P_o + k_2 P)\Delta l \qquad 7.4$$

where $-Q_s$ is the extra heat liberated on shortening a distance Δl, under a load of P g/cm², for a muscle developing isometric tension, P_o.

Since the load on the muscle determines the velocity at which it will shorten (cf. Hill's equation 6.1, or Aubert's 6.2) it follows that the extra heat per cm of shortening is also related to the velocity of shortening, being small at high velocities and increasing rapidly as the velocity decreases.

The constants k_1 and k_2 in equation 7.4 are dimensionless. Average values for them are $3{\cdot}75 \times 10^{-3}$ and $4{\cdot}2 \times 10^{-3}$, if the extra heat is expressed in mcal/g. Both constants have a coefficient of variation of at least ±30 per cent, so that the terms in equation 7.4 can be reduced to $K(P_o + P)$, without losing much accuracy. On average,

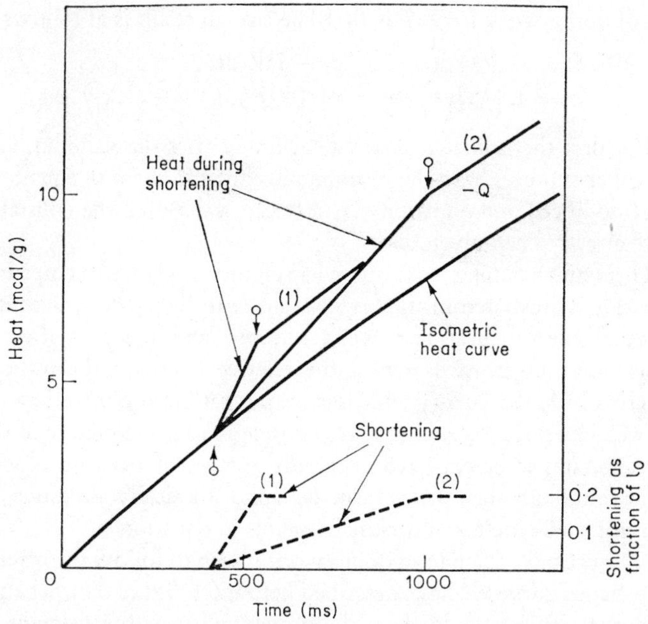

Figure 7.5. The effect on the heat output of release of a muscle from isometric restraint during a tetanus. Release = $0{\cdot}2\ l_o$ (for method of calculation see text). $P_o = 2000$ g/cm²; $V_m = 2$ muscle lengths per sec. Frog sartorius at 0°C.

Release at 0·4 sec after shock (arrow ♂):

(1) under load of $0{\cdot}05\ P_o = 100$ g/cm².

(2) under load of $0{\cdot}50\ P_o = 100$ g/cm².

Upper curves – heat; lower curves (dotted) – shortening.

Shortening comes to an end at arrow (♀).

Calculated after [67].

K then $= 4 \times 10^{-3}$. For a discussion of these values and also for a description of how the curves in figure 7.5 were derived, see [74].

If we wish to calculate the *total* rate of heat and energy production during the time course of a release experiment, it is first necessary to add the isometric heat rate to the extra, shortening heat rate, calculated from 7.4. This can be done by using Aubert's equation for the heat rate during an isometric tetanus (7.3). To arrive at the rate of total energy output, we must then add in the rate of doing work ($= -P\mathrm{d}l/\mathrm{d}t$). The overall result is as follows:

$$-\mathrm{d}E/\mathrm{d}t = -\mathrm{d}Q_s/\mathrm{d}t - \mathrm{d}Q_i/\mathrm{d}t - P\mathrm{d}l/\mathrm{d}t \qquad 7{\cdot}5$$
$$= -[K(P_o + P) + P]\,\mathrm{d}l/\mathrm{d}t + h_A \exp(-\alpha t) + h_B$$

$-\mathrm{d}Q_i/\mathrm{d}t$ is the isometric heat rate at time t after the stimulus, and the other symbols have the meanings assigned to them in equations 7.4 (modified) and equation 7.3. Average values for the constants have already been given.

The result of using this somewhat clumsy and highly empirical equation is illustrated in figures 7.6*a* and *b*. In figure 7.6*a*, the rates of total energy, heat and work outputs, and the velocities of shortening, are plotted against the relative loads on the muscle, whereas in *b*, the energy and heat output rates are plotted against the velocities of shortening. These curves give a good picture of the average rates to be expected in the early stages of a release experiment, although their shapes can be quite drastically modified, if extreme value are taken for the constants in equation 7.4.

In figure 7.6*a*, the rate of doing work is seen to follow the typical bell-shaped curve we have described before (cf. figure 6.8), when it is plotted against the load on the muscle, whereas the velocity of shortening falls in a characteristically exponential manner. The heat rate also falls exponentially from a high value at zero load (=maximal velocity of shortening) to about $\frac{1}{3}$ this value at P_o, that is, under isometric conditions. The curve for total energy rate, on the other hand, is slightly bell-shaped, rising to a maximum at relative loads between 0·1 and 0·2 P_o, and then falling steadily away to about $\frac{1}{3}$ this value at P_o.

When the energy and heat rates are plotted against the velocity of shortening, as in *b*, the energy rate rises sharply from the isometric value at zero velocity to a maximum when the velocity is between 50 and 60 per cent max., and then declines slightly again, until maximal velocity is reached. The heat rate, on the other hand, rises steadily from the isometric value at zero velocity to a maximum at maximal velocity, the curve being nearly linear for a large part of its course. The form of the energy curve might be expected on the sliding filament hypothesis, where movements of the myosin heads

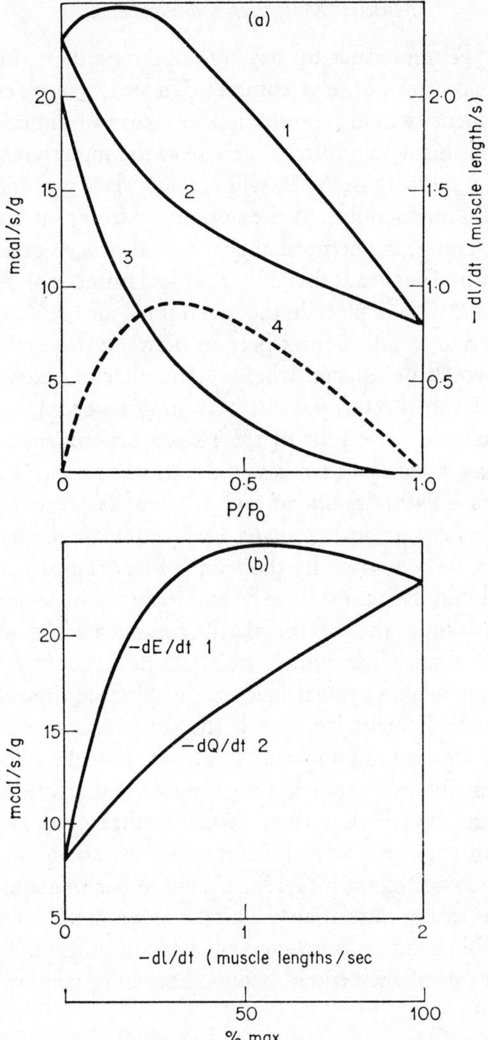

Figure 7.6. (*a*) The rates of energy output during a tetanus of frog sartorius at 0°C, plotted against the relative load on the muscle (P/P_0). Curves calculated from equations 6.1 and 7.5; they give initial rates, at $t \simeq 0$.

Curve 1 $-dE/dt$ (total energy rate)
2 $-dQ/dt$ (total heat rate)
3 $-dl/dt$ (shortening rate)
4 dW/dt (work rate)

(*b*) The rate of total energy output ($-dE/dt$) and of heat ($-dQ/dt$) plotted against the rate of shortening ($-dl/dt$).

are taken to be responsible for pulling or pushing the actin filaments towards the centres of the sarcomeres. Under isometric conditions, such movements would be restricted to a kind of dithering to and fro [31], and this would restrict the rate of the underlying chemical reaction, the splitting of ATP, which takes place as a result of the myosin/actin interactions. At the opposite extreme, at high velocities of shortening, the actin filaments would pass so quickly along the chain of myosin heads that only a limited number of interactions would be able to take place in the given time, and this would cause the energy rate to fall. Somewhere in between these extremes, an optimum would be reached where the number of interactions was maximal. We shall return later to this important topic.

To summarize the results of the release experiments: *extra heat is given out over and above the isometric heat, when a muscle is suddenly released from isometric constraint and this heat is proportional to the shortening (at constant load) and to the load (at constant shortening)*. This law deserves to be known by the name of its originator A. V. Hill [70, 71, 74]. Similar results have been obtained with single twitches but the situation is then unnecessarily complicated by the limited duration of the twitch which restricts the amount of external shortening under the heavier loads because of the time needed for the contractile elements to stretch the series elastic elements (see figure 6.3). Under light and medium loads, however, heat appears to be given out in proportion to the shortening during limited phases of the twitch, but there is the further complication that evolution of the early activation heat is then not quite complete before shortening begins; it therefore tends to become confused with the true shortening heat, if any. There is also a definite tendency in all the published curves for the rate of heat output to fall to zero in the later phases of the twitch, before shortening is quite complete [71].

At this stage, it is only fair to warn the reader that the evolution of shortening heat, although apparently quite clear cut in the isotonic release experiments we have just described, is still a highly disputed question, particularly when the underlying chemical reaction, the splitting of ATP, is considered in relation to it. As we shall see in the next chapter, none of the measurements of ATP splitting during a living contraction has so far shown there to be any chemical equivalent of this shortening heat. To overcome the apparent contradiction between the two sorts of measurement, it has been suggested that there might be a phase of *heat uptake* by the muscle during relaxation [29, 30, 31], but this has certainly not been demonstrated experimentally, either in the isotonic tetanus or the twitch. A similar apparent contradiction is also brought out amongst the heat measurements themselves, when the measured heat output during the rising, contractile phase of a twitch is compared with that for the overall cycle of contraction and relaxation under a falling load, as we shall now see.

Energy output in isotonic twitches

The energy output of a twitch, when divided up into its two terms, heat and work, is highly dependent on the load on the muscle, which also determines the velocity of shortening, as we have seen. There is, however, a marked difference in the form of the experimental curves, depending on whether the whole cycle under a falling load is considered or only the contractile phase. It is simplest to discuss the whole cycle first (figure 7.7*a*. full lines): here the total heat given out in a series of twitches was collected up by a slow thermopile, and according to equation 7.2, represents the total energy output, because the load fell back to its starting point during relaxation [24]. The work done during the contractile phase was also measured, so that by substraction of the latter from the former parameter, we have a measure of the heat output ($-Q$), due to processes other than the degradation of work into heat as the load falls during relaxation. These three parameters, $-\Delta E$, W and $-Q$ are plotted in the figure against the relative load on the muscle (P/P_0).

We see that the work curve (3*b*) is of the typical bell-shaped form, also shown by the work output during a tetanus (figure 7.6). The total energy output (1*b*) has an exactly similar form, but the intercepts on the ordinate, instead of being zero, are positive (at zero load, $-\Delta E =$ about 2·5 mcal/g, and at $P = P_0$, it equals about

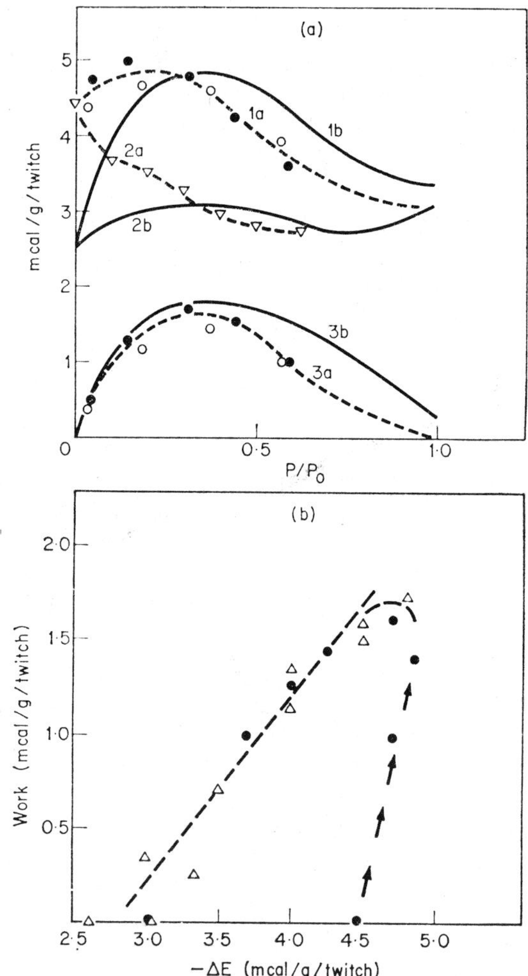

Figure 7.7. (*a*) Overall energy balance during a twitch, plotted against the relative load (P/P_o). The full lines are averages from [24], for the whole cycle of frog sartorius twitches at 0°C, with a falling load during relaxation (curves 1*b*, 2*b* and 3*b*). The dotted curves are from [71], for twitches of frog sartorius (circles) and toad semi-membranosus (dots), during the rising contractile phase of the twitch only (1*a*, 2*a* and 3*a*). $-\Delta E$ = total energy output, Q = heat output, W = work done in contractile phase of both series.)

(*b*) Plot of the work done against the total energy output, for the experiments in (a).

△ — results from whole cycle with falling load [24].

● — results for rising phase of twitch only [71].

Arrows indicate that the relative load [P/P_o] was increasing from 0 to 0·30 (rising phase of twitch only).

3·1 mcal/g). The difference between the total energy and the work represents $-Q$ (2*b*). We see that this parameter is almost independent of load, although the curve is actually of a shallow S-shape, the values given by it varying between about 2·5 mcal/g and 3·1: this may be a real variation, but more experiments would be needed to prove the point, because of the relatively high standard errors of the measurements. The average value for $-Q = 2{\cdot}9$ mcal/g, independent of load.

The conclusion from this elaborate and carefully controlled series of experiments on the overall cycle [24] is quite clear cut: shortening of the muscle, which amounted to at least 35 per cent l_o at zero load, but to less than 5 per cent at $P = P_o$, was obviously *not* accompanied by any extra output of heat ($-Q$): this is particularly clear at zero load, where the heat output was *smaller* than at any other loading, but the amount of shortening was *greater*. Hence, we can write a very simple approximate equation for the energy production under these conditions:

$$\begin{aligned} -\Delta E &= \text{constant} + \text{kW} \qquad 7.6 \\ &= 2{\cdot}9 + 0{\cdot}97 \text{ W (in mcal/g/twitch)} \end{aligned}$$

We now turn to an earlier series of experiments [71] on the contractile phase of frog and toad twitches. This series was unmistakably claimed at the time to demonstrate that shortening of the muscle was always accompanied by a proportionate increase in the output of heat. Indeed, the *experimental* curves given in figure 7.7 certainly do seem to show a marked effect of shortening, because now the heat output, $-Q$, is highest at zero load where shortening was greatest, and falls steeply as the load on the muscle is increased (curve 2*a*). Moreover, the value at zero load is 1·7 mcal/g higher than that for the overall cycle which was 2·5 mcal/g. This large difference in the heat output between the two sets of conditions is reflected, of course, in an equally great difference in total energy output, $-\Delta E$, which follows a far less markedly bell-shaped course for the contractile phase (1*a*) than for the overall cycle (1*b*). Note that the work curves are of the same bell-shape in both cases; the reason for the differences in absolute amount are probably due to the fact that the internal work done against the series elastic components was taken into account in the experiments on the overall cycle, but not in those for the contractile phase.

Another way of bringing out the differences between the two series of experiments is shown in figure 7.7*b*, where the work done is plotted on the ordinate and the total energy output on the abscissa. We see that the points for the overall cycle (triangles) lie close to a single straight line, as would be expected from equation 7.6; the line we have drawn cuts the abscissa at $-\Delta E = \sim 2{\cdot}8$ mcal/g. The points for the contractile phase only (dots) also follow this straight line in the higher range of loads from 0·3 P_o up to P_o, but in the lower range of loads from $P = 0$ to 0·3 P_o, they lie far to the right, on a curve of their own. This could be interpreted to mean that shortening, when it exceeds the value characteristic for a load of 0·3 P_o, is increasingly accompanied by extra heat, over and above that found for the whole cycle; this conclusion runs quite counter to equation 7.6 [165].

There are several ways of explaining the discrepancies, in the lower range of loads, between the contractile phase and the overall cycle; for example, the idea that there is a phase of heat uptake during relaxation which exactly balances the extra heat given out in the contractile phase; but as we noted above, there is at present no experimental evidence that this is so. It might even turn out that large experimental errors were introduced into the heat measurements, when the muscle shortened very rapidly, although both sets of experiments were carefully controlled, precisely to obviate such errors. Whatever the case may be, it does not seem fruitful to balk these difficulties, even though the chemical studies we shall discuss support the results of the heat measurements obtained for the overall cycle and, therefore, contradict the earlier results for the contractile phase only. Perhaps it might be justified to ignore the discrepancies, if we had to rely solely on isotonic twitches to support the concept that extra heat accompanies shortening, but we have already seen that such heat appears even more unmistakably when the muscle is suddenly released to a new length during a tetanus (figure 7.5 and equations 7.4 and 7.5). There is, however, one loophole: that is, that the curves in figure 7.5 for the release experiments are calculated ones, and therefore highly idealized; in practice, the release of extra heat often follows the shortening only for a limited part of its time course, and then its rate falls off, so that it more and more closely approaches the 'isometric' heat curve. At the end of the release, the result is, therefore, a much smaller difference in the total amounts of heat, given out in the two cases, than equation 7.4 would suggest.

Returning finally to the experiments on the whole cycle, we can restate equation 7.6 in the form: *the energy given out in the whole cycle of a twitch, with a falling load during relaxation, is proportional to the work done plus a constant amount of heat, which arises during the transduction of chemical energy into work.* This conclusion goes back historically a long way to the work of Fenn [47], who first stated it in 1923; note that Fenn was then studying the isotonic tetanus and not the twitch, so that this law, which is known by his name, can evidently be applied to both these types of experiment. In that case, the discrepancies we have spoken of can be most clearly brought out by comparing Fenn's law with Hill's (p. 154).

Efficiency of the contractile process

The mechanical efficiency of contraction is given by $-W/\Delta E$ ($= W/W - Q$), and this indeed is the only measure of efficiency which we have, since to estimate the thermodynamic efficiency, $-W/\Delta F$, a knowledge of the free energy changes ($-\Delta F$) is required, and these are still uncertain [163]. We can obtain two sets of values for $-W/\Delta E$ from figure 7.7*a*, one for the whole cycle of the twitch [24], and the other for the contractile phase [71]. These are shown respectively by a broken and a full line in figure 7.8, for loads from 0 to 0·6 P_o, where the efficiency has just passed its maximum. From this point on, the efficiency declines to zero at P_o, if the external work only is considered, but to a higher value of at least 8 per cent, if internal work is also included in the energy balance. Unfortunately very few values are available for the intermediate zone between 0·6 P_o and P_o, where it is also difficult to estimate the internal work.

The efficiency up to 0·3 P/P_0 is higher for the whole cycle than for the contractile phase, which is a necessary consequence of the greater amount of heat, but similar amount of work given out in the contractile phase, in this range of loads (cf. figure 7.7*a*). The maximum efficiency is about 36 per cent in both cases, and this

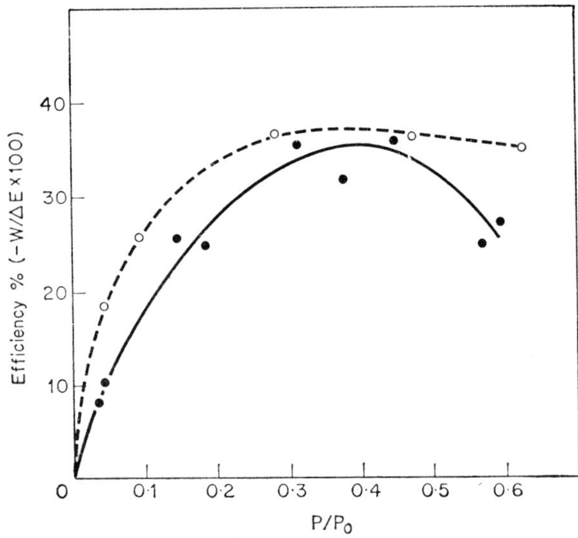

Figure 7.8. The mechanical efficiency ($-W/\triangle E$) for the rising phase of frog and toad twitches (full line and dots), and for the whole cycle, with falling load, of frog twitches (broken line). Values taken from previous figures, and dotted against P/P_0.

maximum extends from about 0·3 to about 0·45 P_0. Beyond this point the efficiency for the contractile phase of the twitch tends to fall off more markedly than that for the whole cycle, probably because the experiments under the highest loads in the former case were not continued for quite long enough to give the maximum work. It is possible to obtain higher maximal efficiencies than 36 per cent by using an ergometer and thus forcing the muscle to work at constant speed [71]. A value of 45 per cent seems to be the upper limit in frog and toad muscle, but in tortoise muscle values of 75 to 80 have recently been recorded (see [165].

Table 7. Early heat output during isometric twitches in various muscles, at 0°C [61, 71–73, 165]

Muscle	*Heat*			*Tension*
	Time of onset after stimulus (m.sec)	*Initial rate* m.cal/g/sec	*Rate at* ½ *max. tension*	*Time of onset* (m.sec)
Frog sartorius	8–10	35·0	<10·0	23
Toad sartorius:				
normal Ringer	15	17·0	<7·0	40
1·86 × R	27	17·0	—	70
2·28 × R	30	14·0	—	150
Tortoise iliotibialis	60	4·0	<1·0	90

(1·86 and 2·28 × R indicate that the strength of the Ringer solutions, initially bathing the muscle, were increased in this proportion)

8: Chemistry of Contraction

On the basis of *in-vitro* experiments described in part I, it was taken for granted that the energy for contraction came directly from the splitting of ATP. It is now necessary to demonstrate that such a splitting of ATP does in fact occur in a living contraction in sufficient quantity to supply the energy. This is a difficult aim to achieve in practice, mainly because of the very small quantities involved; for example, the total energy output during a twitch at maximal efficiency is 5 mcal/g muscle, and taking the total energy change as 10 kcal/mole ATP, this would amount to a splitting of only 0·5 μmoles ATP/g/twitch. In frog sartorius muscle, the initial ATP content is about 3 μmoles/g, but it can vary to at least ±0·3 μmoles from muscle to muscle in a pair; this introduces a considerable error, since the only possible way of estimating the change is to compare pairs of muscles from the same animal, a control muscle with a stimulated muscle [29].

There is in addition the problem of how to arrest the reactions as instantaneously as possible; the usual method is to freeze the muscle, before extracting it with perchloric or trichloracetic acid. This is now done by plunging the muscle, still in position on the recording apparatus, into a flask of liquid freon ($CCl_2F_2 + CCl_3F$) at −180°C, by raising the flask rapidly from below [86, 87, 29]; it is then possible to freeze the muscle within 200 msec, and in some methods even within 50 msec. The technique avoids one of the snags inherent in the older method of freezing directly in liquid nitrogen, where a layer of vapour tended to form between the muscle and the liquid, so insulating the muscle and preventing rapid freezing.

Having overcome the problem of arresting the changes, we must next consider which reactions, in the whole chain of resynthesis of

ATP, can occur in the given time interval. Fortunately this problem is not as difficult to solve as the others. The main subsidiary reaction which occurs with the greatest rapidity is the so-called Lohmann reaction (*b* below), catalysed by creatine kinase:

$$\left.\begin{array}{ll}(a) & \text{ATP} \underset{(\leftarrow)}{\rightleftarrows} \text{ADP} + \text{P}_1 \quad (\text{ATP-ase}) \\ (b) & \text{ADP} + \text{PC} \rightleftarrows \text{ATP} + \text{creatine (Cr)}\end{array}\right\} \qquad 8.1$$

Since the initial content of PC is about 20 μmoles/g, and the maximum amount broken down in a long series of twitches rarely exceeds 5 μmoles/g, the equilibrium in practice lies *to the right*, though not so strongly as is often supposed. Hence the amount of PC hydrolysed is a good measure of the ATP broken down, providing the experiment lasts long enough for completion of the reaction, but not so long as to allow other side reactions time to occur; the limit in this respect is about 3 sec at 0°C [88].

The time course of the Lohmann reaction can be followed during the twitch by the pH changes which occur in the muscle [32]. There are two ways of measuring these changes: either to place a micro-glass electrode on the surface of the muscle, or to ingest a pH indicator into the living muscles by the technique used for murexide ingestion, and follow the colour change photometrically (cf. chapter 4). The basis of the methods is that an acidification of the muscle takes place when ATP is split, according to equation 2.2 [53]. This amounts to 0·8 to 1·0 proton equivalents per mole ATP split at physiological pH values. The Lohmann reaction, on the other hand, causes an alkalization amounting to −1·0 proton equivalents per mole PC [32]. If the latter reaction is not quite instantaneous, but lags more or less behind the initial splitting reaction, 8.1*a*, the pH will first fall and then rise again, back to its initial value. This is exactly what occurs in a contraction: the acidification from ATP splitting occurs during the rising phase of the twitch, whereas the alkalization due to the Lohmann reaction does not set in until relaxation begins and often continues for 100 to 200 msec after it is complete. From this and from the results of a series of superimposed twitches, where the alkaline phase only becomes apparent during the last twitch of the series, it is clear that the Lohmann reaction is slower than the splitting reaction [32]. For this reason, it is very important in contraction studies to allow relaxation time to occur before freezing the muscle, if the changes in PC or in creatine are used as the criterion of the breakdown of ATP. Unfortunately, it is by no means clear from the literature that this has always been done.

When relaxation is complete, the reactions of the glycolytic cycle set in; these start with the phosphorylation of glycogen by the inorganic phosphate (P_i) released in reaction 8.1*a*, and involve, amongst other things, the dephosphorylation of 1 molecule of ATP per glucose unit of glycogen, as fructose diphosphate is formed. Luckily, most of the complications due to this cycle can be obviated by poisoning the muscle with iodoacetate [23, 24, 104, 107], which blocks glycolysis at the triose phosphate dehydrogenase stage. This enables accurate measurements of PC disappearance, or its equivalent, the appearance of free creatine, to be made after a long series of twitches or a long tetanus, with a reasonable assurance that only reactions 8.1*a* and *b* have taken place. By these means, the relations of PC breakdown to the energy output during contraction can be established.

To go further and estimate ATP breakdown itself proved impossible until quite recently, when it was discovered that fluorodinitrobenzene (FDNB), the reagent used extensively for amino end-group analysis of proteins, was to some extent a specific poison for creatine kinase. In this way, reaction 8.1*b* can be inhibited without doing too much damage to the contractile machine [88, 100]. This made it possible to show that ATP was indeed broken down, in single twitches or tetani, in amounts corresponding to the breakdown of PC in the presence of iodoacetate as inhibitor, or to the amount of P_i liberated in short 'unpoisoned' contractions. Even here, another reaction tends to complicate the results; this is the myokinase reaction:

$$2\text{ADP} \rightleftharpoons \text{ATP} + \text{AMP} \qquad 8.2$$

This, in its turn, leads to deamination of the AMP, catalysed by adenosine monophosphate deaminase:

$$\text{AMP} \longrightarrow \text{inosine monophosphate} + \text{NH}_3 \qquad 8.3$$

These two reactions intervene to a significant extent both in frog sartorius and rectus abdominis muscles, poisoned with FDNB, so that corrections for them have accordingly to be made [29], if $-\Delta$ATP is measured, but not if ΔP_i is used to assess the change.

The outcome is that we now know for certain that ATP is broken down during contraction, and that this reaction is the closest to the contractile process.

Breakdown of ATP and PC during short isometric tetani

The time course of the breakdown of ATP and the release of extra P_i during short isometric tetani of 1·5 sec duration, in FDNB-poisoned frog sartorii at 0°C, is shown in figure 8.1 [87]. Also shown are the development of tension, and the output of heat; the latter was calculated according to Aubert's equation, using the same constants as in figure 7.5 and equation 7.3, after correction for the effect of FDNB. It is quite remarkable how well the curves for $-\Delta$ATP and heat output agree, when particular account is taken of the fact that the steady rate of heat output from FDNB-poisoned muscles is usually only 70 per cent of that from untreated muscles [4]. It is also seen that the breakdown of ATP (dots) agrees well with the release of P_i (circles); this shows that the arguments in the introduction to this chapter are justified.

We can make a rough calculation of the enthalpy of hydrolysis of ATP by comparing the amount of heat released ($-Q$) with the amount of breakdown ($-\Delta$ATP), about halfway through the tetanus at 0·7 sec. These are respectively 4·70 mcal/g and 0·46 μmoles/g, giving an enthalpy of hydrolysis (dE/dATP) of 10·2 kcal/mole ATP. We may compare this value with that obtained in a series of experiments on very long tetani, lasting from 20 to 60 sec [106, 107]. Here the average rate of ATP breakdown, measured in this case as PC breakdown, was very much lower, in line with the lower heat rate in long tetani (cf. equation 7.2 et seq.); it amounted in one series of experiments to 0·28 μmoles/g/sec and in another to 0·23 μmoles. A mean value for the stable heat rate (h_B) from [3] is 3·5 mcal/sec/g, thus giving a rather lower value for dE/dATP of only 7·3 kcal/mole.

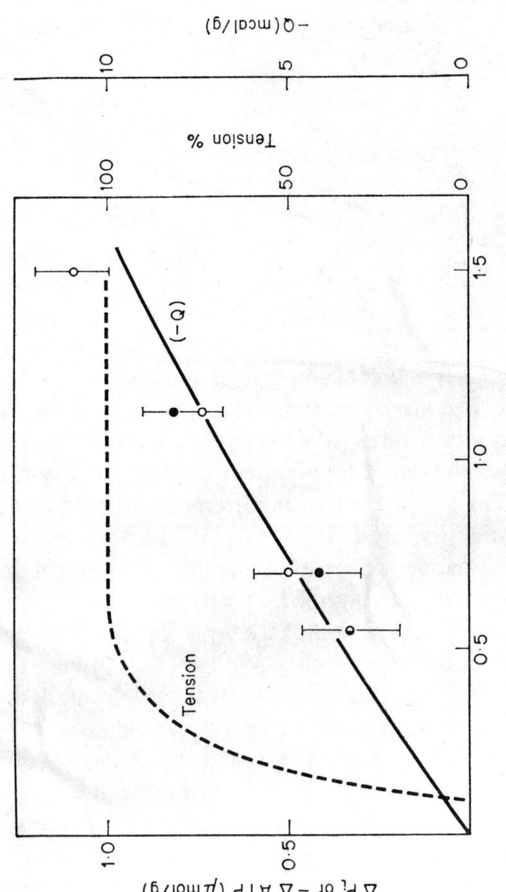

Figure 8.1. The breakdown of ATP (●) and the appearance of Pi (○) during a short isometric tetanus of frog sartorius muscle at 0°C. After [87]. Muscles were poisoned with FDNB to inhibit PC-kinase.

Vertical bars represent the SE's of the chemical estimations.

The line for the heat output (−Q) is calculated from Aubert's heat equation (7.2), after reduction of its second constant (h_B) to 70 per cent of its normal value, to allow for effect of FDNB (see [4]).

Effect of muscle length on the rate of breakdown of ATP during isometric tetani

Like the heat rate (cf. figure 7.4), the rate of breakdown of ATP during an isometric tetanus is dependent on muscle length. This is illustrated in figure 8.2 from two series of measurements by

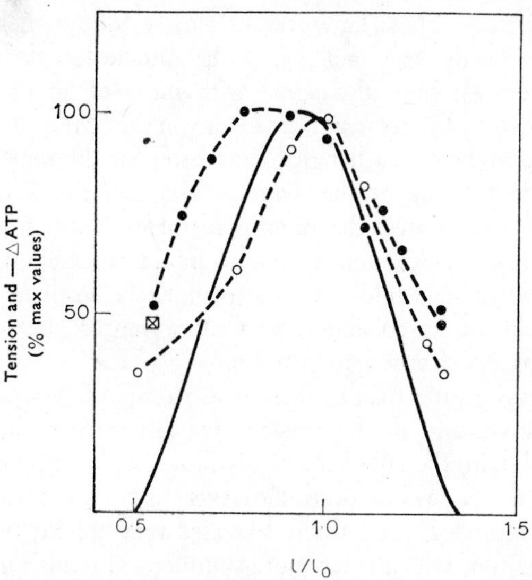

Figure 8.2. The effect of muscle length on the breakdown of ATP (measured as △Pi or as —△PC) during tetani of frog sartorius muscles at 0°C. The full line represents the development of tension, and the dotted lines the appearance of Pi (circles) or breakdown of PC (dots). The two extra points (a cross and a square) at 0·5 l/l_0 are from [106], whereas the circles are from [87], and the dots from [134]. Muscle length adjusted to make the tension diagrams, given by the various authors, agree.

different investigators [87, 134]. In the first series (circles) a 1·5 sec tetanus was used, and the breakdown was measured as P_i [87], whereas in the second (dots) the tetanus was of 20 sec duration, and the chemical change was measured as —ΔPC [134]. The length/tension diagrams of both series were nearly identical in shape, and could be made to coincide by adjusting the relative muscle length by 0·1 l/l_0 from one series to the other; they are shown as a common line.

We see from the figure that tension and breakdown fall hand in hand as the length is reduced, in the case of the 1·5 sec tetani (circles), except at $l/l_0 = 0{\cdot}5$, where the tension has fallen to zero, but the relative breakdown of ATP is still 35 per cent of maximal. On the other hand, in the longer tetani (dots), the relative breakdown is always greater than the relative tension on this side of the diagram. These latter results closely follow those for the fall in the steady heat rate, h_B, as the muscle length is reduced (see figure 7.4); they also agree with those of another investigator, shown by the cross and the square on the figure, at $l/l_0 = 0{\cdot}5$ [106]. We suggested earlier that the reason for the rapid decay of tension on this side of the diagram was that the Z-discs were beginning to compress the myosin filaments (as in figure 6.7*b*); this compression, although tending to inhibit interactions between myosin and actin, would not do so completely, so that splitting of ATP would still be possible at very short lengths: this is precisely what figure 8.2 clearly demonstrates.

At lengths greater than l_0, there is also a rapid fall in the amount of breakdown and in the tension; the fall in breakdown again agrees well with the fall of the steady heat rate, h_B, shown in figure 7.4 under the same conditions. However, in both series of experiments in figure 8.2, the tension decreased very quickly on this side of the diagram, which is far more symmetrical than is usually the case (cf. figure 7.4). For this reason, the results leave us in some doubt about how ATP breakdown is quantitatively related to tension, although they show clearly that it behaves in relation to length, as would be expected on the sliding filament theory: that is, it falls off as actin is pulled out of contact with myosin at increasing lengths.

Breakdown of PC or ATP during isotonic twitches and tetani

The breakdown of ATP during single isotonic twitches and tetani has been investigated mainly by three schools of workers, each of whom have a rather different version of what happens. The earliest reliable investigations were carried out on frog sartorius muscles at 0°C, using either a series of twitches, or short tetani, under varying load, and measuring the breakdown of ATP in terms of the accumulation of free creatine [113]. All the muscles were poisoned with iodoacetate. The results showed definitely that for a single isotonic tetanus of 400 m.sec duration at 0°C, the appearance of free creatine (Cr) followed an equation of the type:

$$\Delta \mathrm{Cr} = A + k'W - k''\Delta l/l_0 \quad 8.4$$

where A is a constant independent of the work done or the velocity of shortening; $1/k'$ can be termed the mechanochemical equivalent ($dW/d\mathrm{Cr}$), and k'' is a term representing the contribution of shortening *per se* to the utilization of energy. It was found that k'' was so variable that its difference from zero was not statistically significant. Hence it was concluded that *there was no term in the energy equation equivalent to the large shortening heat of Hill* (cf. figure 7.5 and equation 7.4). The constant, k', had a value of 0·11, with work in mcal and ΔCr in μmoles/g, thus giving a value for $dW/d\mathrm{Cr}$ of 9 kcal/mole ($= -dW/d\mathrm{PC} = -dW/d\mathrm{ATP}$).

Three other similar series of experiments, but this time comparing pairs of muscles, one contracting isotonically and the other isometrically, gave values for $-dW/d\mathrm{PC}$ or $dW/d\mathrm{Cr}$ varying from 7 to 11 kcal/mole. Again there was no term, corresponding to shortening heat. It was not possible in these experiments to obtain a value for the enthalpy of breakdown of PC or ATP, because heat measurements were not included [26, 106, 107].

Fortunately, a second school carried out a very carefully controlled series of experiments on the breakdown of PC, in connection with the heat measurements shown in figures 7.7*a* and *b* [24]. The authors compared the total energy output with the breakdown of PC, and found a value for the enthalpy (dE/dPC) of 9.16 kcal/mole for thirty lightly loaded twitches, where very little work was done, but the muscle shortened a long way; of 10·43 for a series of twenty-five more heavily loaded twitches, where there was only a moderate shortening, but large work output; and of 9·58 for thirty isometric twitches, where almost no shortening occurred and again little work was done. Later experiments have revised the latter figure upwards to 10·06 for 10 sec tetani, to 10·22 for 20 sec tetani and to 11·94 for isometric twitches [25]. Hence the best overall figure for the enthalpy or internal energy change from all these experiments is 10·6 kcal/mole. It is this value we shall use henceforth for converting PC or ATP breakdown into total energy change, or vice-versa but see [24], later reference.

The above series of experiments [24] was carried out by comparing pairs of muscles, one contracting isotonically under varying load and the other isometrically, and measuring the difference in PC content between them at the end of a series of twitches. The results are illustrated in figure 8.3.

We see from the figure that at zero load, and hence zero work, the difference in breakdown of PC between isotonic and isometric conditions is negative, *that is to say* 0·08 *μmoles less PC are broken down per g when the muscle shortens, but does no work, than when it contracts isometrically*. As the muscle is loaded increasingly, however, more PC begins to be broken down than under isometric conditions, and this increased breakdown goes hand in hand with the increased work, until the load = 0·3 P/P_o. On increasing the load beyond this point, the trend is reversed, that is, the breakdown of PC falls off more rapidly than the work; this means that on this side of the diagram there is no longer a strictly linear relation between PC breakdown and work done.

We should also note the small amounts of work done by the poisoned muscles, seen by comparing figure 8.3 with figure 7.7 for unpoisoned muscles. This might well account for the low mechanochemical equivalent in these experiments (see below), and also in those in which FDNB was used to poison the muscles (figures 8.4*a* and *b*). It could similarly explain the serious discrepancy between the constants in the energy equation, 7.6, and those in the 'chemical' equation, 8.5 (which must be multiplied throughout by 10·6 to give $-\Delta E$).

If we accept the straight line relationship discussed above as best representing the relation of PC breakdown to work for loads below 0·3 P/P_0, then its slope ($-\mathrm{d}W/\mathrm{dPC}$) is equivalent to 4·65 kcal/mole PC, a considerably lower mechanochemical equivalent than those

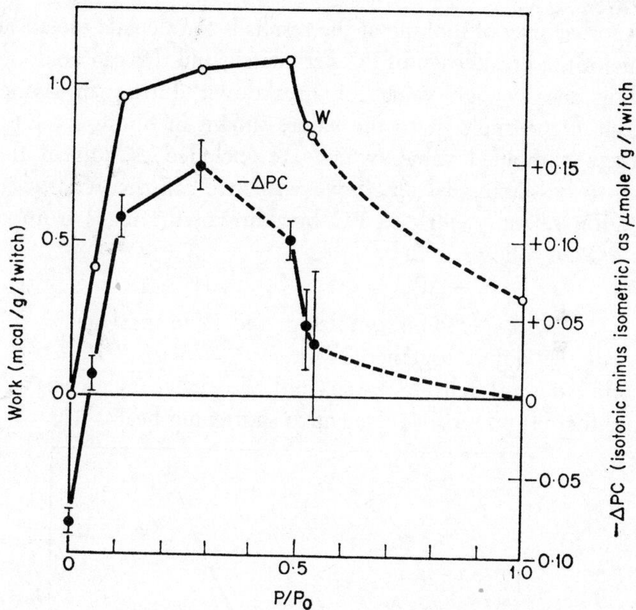

Figure 8.3. The difference in breakdown of PC between isotonic and isometric twitches (isotonic minus isometric) compared with the total work done (*W*).

The muscles were poisoned with iodoacetate before start of experiment. Results plotted against the relative load on the muscles (P/P_o); vertical bars represent SE's of the means of the chemical estimations. Note that the difference in PC breakdown (isotonic minus isometric) at zero load becomes negative; i.e. less PC is broken down in an unloaded isotonic twitch than in an isometric twitch. Results taken from [24].

on pp. 165 and 169. The authors themselves calculated the mean slope for all the points available, which gave a value of 6·00 kcal/mole.

Another way of looking at the results is to calculate the absolute value of the breakdown of PC per twitch, and this can be done by adding the average value for breakdown during an isometric twitch, 0·29 μmole/g, to the values shown in figure 8.3 [26]. It is these combined values which are included as dots in figure 8.4*a*, to be discussed later. If we wish, we can then construct an equation relating work to PC breakdown, for loads from 0 to 0·3 P/P_0 as follows:

$$-\Delta PC = 0{\cdot}21 + 0{\cdot}215\ W \qquad 8.5$$

with $-\Delta PC$ in μmoles/g, and W in mcal/g.

For all the points together, from $P/P_0 = 0$ to $P/P_0 = 1{\cdot}0$, the second constant falls to 0·166, and the first rises to 0·235 [24]. Note there is no term equivalent to shortening heat.

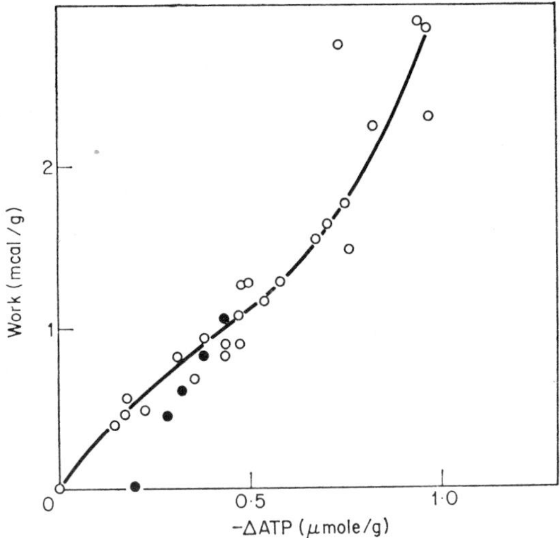

Figure 8.4. (a) Breakdown of ATP or PC during single twitches or short tetani of frog rectus abdominis muscles, plotted against the work done (points, as circles, taken from published results in [23, 86, 87 and 100]). Also included for comparison as dots are results from figure 8.3, after adding 0·29 μmoles of PC breakdown to all values; this is the average PC breakdown during an isometric twitch.

We now come to an extensive series of experiments by a third school, with somewhat different results [23, 86, 87, 88, 100]. Collected values for the breakdown of ATP during a series of twitches or short isotonic tetani from this school are shown in figures 8.4*a* and *b*. The authors compared the contracting muscle with its resting companion from the same animal, and froze the muscles after a single twitch or tetanus in each case. They obtained almost the same results, whether they measured the ATP breakdown directly in FDNB-poisoned muscles or as PC breakdown in iodoacetate-poisoned muscles or P_i appearance in unpoisoned muscles, so that no distinction is made between the methods in the figure. It should be noted that it was this school which first definitely

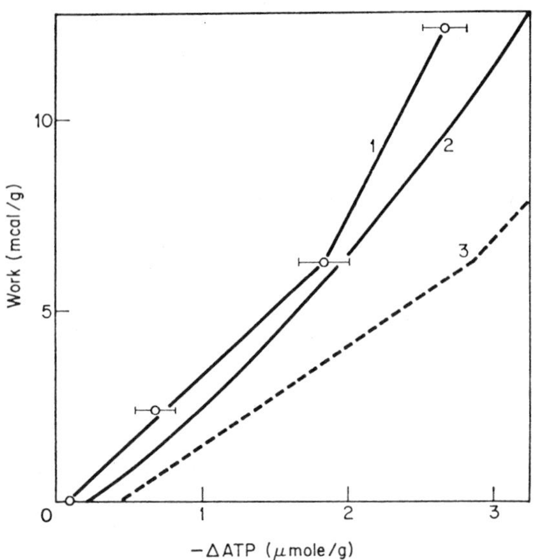

(*b*) ATP breakdown during twitches and long and short tetani of frog sartorius muscles at 0°C (FDNB poisoned), compared with the work done. Note low mechano-chemical equivalent of 3·4 to 5·0 kcal/mole.

Curve 1 – ATP breakdown from [100] with SE's of means of ATP values (horizontal bars).

2 – Predicted from Aubert's heat equation, 7.3, for tetani (*no* shortening heat).

3 – Predicted from Aubert's equation with addition of values for shortening heat, calculated from Hill's equation (7.4).

showed that ATP itself did indeed break down during a contraction, and thus successfully took up a famous challenge to biochemists to demonstrate this vital fact [cf. 67].

Three points are immediately apparent from the figures: (i) ATP breakdown during contraction of frog rectus abdominis muscles at 0°C (figure 8.4*a*) plots linearly against the work done, up to work values of about 1·8 mcal/g; the line can be extrapolated back nearly through the origin: (ii) the incomplete results for frog sartorius in figure 8.4*b* are similar to those for the rectus, but both sets give very low values for the mechano-chemical equivalent (2·2 to 5·0 kcal/mole ATP; (iii) the results differ from those we have just discussed (figure 8.3), in that the latter predict a breakdown of about 0·2 μmoles of ATP/g when no work is done, whereas the former show almost zero ATP breakdown under these conditions. At higher values of the work, however, both sets of results agree fairly well (dots and circles in figure 8.4*a*).

Recently, careful reinvestigation of lightly loaded twitches of frog sartorius, by these authors [30], have confirmed that there is very little breakdown of ATP when no work is done under isotonic conditions, in spite of the fact that the muscles can shorten a long way. The average breakdown for a single unloaded twitch was found to be 0·08 $\pm$ 0·0009 μmoles/g.

To confirm this result and to show that ATP breakdown can in no way account for the so-called shortening heat, a further series of experiments was carried out in which one member of a muscle pair contracted fully under zero load, and the other was restrained under isometric conditions. If shortening heat had any equivalent in terms of ATP breakdown, it can be calculated from equation 7.4 that the freely shortening muscles should have given out 2·3 mcal of extra heat, equivalent to breakdown of 0·22 μmoles ATP/g, over and above that under isometric constraint. In fact, the result was exactly the opposite: the freely shortening muscles broke down 0·128 μmoles *less* ATP than those under isometric conditions. This is not very different from the results in figure 8.3*a*, where 0·08 μmoles *less* ATP were broken down under zero load than under isometric constraint. It would yield a value of 0·29–0·128 = 0·162 μmoles for the ATP broken down per g per twitch, when the muscle is allowed

to shorten freely. The various values for the breakdown of ATP or PC during unloaded contractions are summarized in table 8.

The absence of any term representing shortening heat is indirectly confirmed by the results for the long tetani of frog sartorius, shown in figure 8.4*b* (upper two points) [100]. These represent respectively 2·5 and 4·9 sec tetani, during which the muscles shortened 2·5 cm and 2·94 cm, under two different loads. If the suppositions of the last chapter are correct, the heat output in these cases should be made up in part by the so-called maintenance heat, which can be calculated according to Aubert's heat equation 7.3, and in part by the extra heat associated with shortening, which can be calculated according to Hill's equation 7.4. Using constants for Aubert's equation, corrected for the effect of FDNB according to the author, we find that the heat of maintenance for the 2·5 sec tetanus should be 12·8 mcal/g, which together with the measured work of 6·3 mcal gives a total energy output of 19·1 mcal. At an enthalpy of 10·6 kcal/mole ATP, this would require the breakdown of 1·80 μmoles ATP/g, compared with the found value of 1·83. Similarly for the 4·9 sec tetanus, the calculated maintenance heat would be 20·3 mcal, the work done was 12·4 mcal, and thus the total energy output would be 32·7 mcal from these two sources, equivalent to a breakdown of ATP of 3·1 μmoles/g. The found value was 2·64 μmoles. The effect of adding on shortening heat is shown in curve 3 of the figure. It would contribute 9·8 mcal in the first case, and 12·5 in the second, to give total energy outputs of 30·1 and 48·7 mcal, that is to a breakdown of 2·84 and 4·6 μmoles of ATP, respectively. It is obvious that these values are much too high, and quite outside the experimental error. *Hence, once again, there is no evidence of any breakdown of ATP equivalent to the extra heat of shortening during a tetanus.*

The above conclusions regarding the chemical equivalent of shortening heat have been confirmed in another way, by measuring the changes which occur in the mitochondrial respiratory chain intermediates, particularly in NADH, as a result of the breakdown of ATP during isotonic or isometric twitches and tetani, and its later resynthesis from ADP [91, 92]. It is possible with very thin muscles such as frog or toad sartorius to follow these changes in NADH and other suitable intermediates, by spectrophotometry, using a double beam instrument. It was found that the amount of ADP appearing as a result of the twitches or tetani was substantially independent of how far the muscle may have shortened [92]. It was also confirmed that a muscle contracting freely, and therefore shortening a great deal, breaks down less ATP than one stimulated under isometric conditions. Because this method does not involve destruction of the muscle the measurements can be repeated many times on the same living specimen, so that of all the techniques available this is one of the most promising for the assessment of overall changes.

Effect of stretching a contracting muscle

When a muscle is forcibly stretched a short distance during an isometric tetanus, most of the work done on the muscle is absorbed by it and does not reappear as heat, if the stretch is sufficiently slow. The exact energy relations are complex, because not only the elastic work done, but also the so-called thermoelastic effect has to be taken into account, and there is consequently some disagreement between the various authorities [1, 75, 106]. From the results of the original work some years ago, it was suggested that the work absorbed during the stretch might be used to drive a chemical reaction backwards; the obvious reaction, in the light of more recent knowledge, is the splitting of ATP.

In a sense, attempts to find a resynthesis of ATP during the stretching of an active muscle are likely to be doomed from the start, because a moment's thought shows that the important product of ATP splitting, ADP, can have only a transient existence in contracting muscle before it is either rephosphorylated via the creatine-kinase reaction or is dismuted by myokinase (equations 8.1*b* and 8.2). Hence, there will be virtually no ADP available to act as substrate for direct ATP resynthesis; even if there were, the reaction would have to take place by the highly unlikely route of reversal of equation 8.1*a*.

Two recent series of experiments have been carried out with the above aim in view, and although they inevitably failed in their object, they did at least show that a slow stretch reduced the amount of ATP which would otherwise have been broken down in the contraction [88, 106]. This ATP-sparing effect can, with care, be taken almost to the limit, where no ATP at all is split during the stretch [88]. This result is not as surprising as it might seem at first sight, because it follows that, if the sliding filament hypothesis is correct, contraction is the result of the transduction of the free energy of ATP-splitting into movements of the myosin heads, which then pull or push the actin filaments towards the centre of each sarcomere; it would therefore be expected that dragging these active heads unwillingly away from their actin partners by a forcible stretch would naturally inhibit the interactions which lead to splitting. These elaborate stretch experiments thus seem merely to confirm what we already know from a wealth of other evidence: the sliding filament theory of contraction explains most of the essential features of active muscle. The problem of what happens to the work absorbed during a slow stretch of active muscle remains, however, unsolved.

Energetics – a summary

The reader can be forgiven if he is now somewhat confused by apparently contradictory findings in the field of energetics. For instance, if he is familiar with the history of the subject, he will have been surprised to discover that the concept of an extra output of heat, specifically associated with shortening, which seemed to have been so well established during the previous two or three decades, has recently suffered some severe shocks at the hands of biochemists and physiologists alike. But this is really only a special example of the quirks of history, because this very concept had itself replaced the earlier hypothesis of Fenn, which had been completely overlooked until quite recently and then was brought back into the limelight and proved to apply to the overall cycles of twitches and short tetani alike. We may well ask the very pertinent question: which set of findings are we to rely upon, since both are claimed to be supported by highly accurate measurements?

Perhaps the answer to the question is that both sets of findings are correct within their own particular limits. Thus, it seems quite certain that more heat is given off while a muscle is actually shortening than when it is held stationary under isometric restraint. This follows from Hill's law and the mass of experimental results which support it [2, 70, 71, 74]. On the other hand, if the muscle is allowed to relax, under a *falling load*, then it is equally clear that the overall energy balance sheet contains no heat term, corresponding to the shortening which occurred in the contractile phase: on the contrary, the energy of the overall cycle is now made up solely of a constant heat term plus a term representing the work which has been done (Fenn's law) [24, 25, 47]. Somewhere between these two sets of findings lie the results obtained by holding up the afterload just as relaxation begins, so that the muscle is not stretched out again [71]. Relaxation is then often formed to be thermally neutral, although sometimes a small uptake of heat can be detected on the heat curves. This means that almost all the extra heat which may have been given out during shortening must remain in the energy balance sheet at the end of relaxation under conditions where the load is not allowed to fall. At first sight, it could indeed be supposed that the above discrepancies were due to an effect of the falling load, such that it could automatically wipe out the extra heat of shortening by causing an equal absorption of heat during relaxation. However, this is not likely to be so, since such a large heat change could hardly have escaped detection by accurate thermopiles [165].

The chemical findings, by whatever method the splitting of ATP is assessed, all agree amongst themselves, and show as conclusively as possible, within the limts of the rather large experimental errors, that ATP is broken down according to Fenn's law and not Hill's: that is to say, it is broken down in proportion to the work done plus a small constant amount of breakdown of the order of 0·20 μmoles/g/twitch, no matter how much the muscle may have shortened during the contractile phase. Even during a tetanus, there is no evidence that extra ATP breakdown is associated with shortening *per se*, but only with the work done. In this case, of course, the constant heat term increases in proportion to the duration of the tetanus, that is, in proportion to the so-called heat of maintenance of the contraction. We can write an approximate equation for ATP breakdown during a tetanus, equivalent to equation 8.5 for a single twitch, by combining the integral form of Aubert's equation for the maintenance heat of a tetanus (7.3) with the known enthalpy of hydrolysis of ATP of 10·6 kcal per mole, as follows:

$$-\Delta \text{ATP} = -\Delta \text{E}/10{\cdot}6$$
$$= [W + h_A/\alpha[1 - \exp(-\alpha t)] + h_B t]/10{\cdot}6 \qquad 8.6$$
(in μmoles/g)

Like equation 8.5, this equation also contains no terms for shortening heat. How well it agrees with experimental findings can be seen from the line in figure 8.4*b*.

Unfortunately we must allow these intriguing questions to rest in this undecided state, because it is quite obvious that the solution to them can only come from experiment and not by inventing further hypotheses which would only add to the confusion.

Table 8. The breakdown of ATP during unloaded isotonic twitches of frog muscles at 0°C, when no work is done. The breakdown during a single isometric twitch is taken as 0·29 μmoles/g, which has been added to obtain values in italics (from [26]).

Muscle	*PC or ATP breakdown (μmoles/g)*		*Ref.*
	Isotonic minus isometric	*Isotonic*	
Rectus abdominis	—	0 ± 0·08	[29]
Sartorius	−0·08* ± 0·007	*0·21*	[24]
	—	0·08 ± 0·009	[30]
	−0·128 ± 0·035	*0·162*	[30]

* Value extrapolated from line in figure 8.3*b*.
N.B.—*Rana pipiens* used in refs. [24] and [29] and *Rana temporaria* in ref. [30].

9: Hindsight and Foresight

In the last eight chapters, we have traversed an enormous field embracing many different disciplines, through which we have been guided by the brilliant advances made by modern cell biology. In keeping to this main pathway, we have inevitably skated over many important topics, vital to a full understanding of contractile mechanism: for example, the detailed structure of actin and myosin, the nature of the enzyme sites, the mechanism of the splitting reaction and the theoretical background to contraction itself. We shall come back to some of these problems in a moment, because it is in the experimental approach to them that the next stage of the advance is likely to be made, and it is here that one would encourage young muscle biochemists to concentrate their research efforts.

There are other areas where further detailed work is required: e.g. in energetics, the vital problem of heat output during the two phases of the contractile cycle and the question whether there really is an endothermic phase during relaxation to balance the apparent heat of shortening during contraction. Similarly, in chemistry, there remains the problem of the mechano-chemical equivalent, which varies widely and is usually much lower than the theoretical value of 11·0 k.cals/mole, deduced by dividing equation 7.6 through by the enthalpy of PC breakdown.

Compared with these deficiencies which are mainly a question of methodology, the advances made by molecular biology and biochemistry have been of colossal proportions. We have only to think of the elegance of the sliding filament theory or the beautiful simplicity of the mechanism for inward transmission of the stimulus, via the sarcoplasmic reticulum; or the equally brilliant discoveries which led to the identification of the triads of the reticulum as a calcium pump, and the linking of this with the known requirement of the enzymic centres on actin and myosin for Ca ions: we can then see how far we have advanced in the last ten years or so from the vague and uncertain position we were in before, when the shifting sands of the unknown region seemed of almost indefinite extent.

There are, indeed, important lessons to be learnt from these discoveries, not least the need for the concentration of many different disciplines on the vital spots of a scientific problem, and for the ability amongst young scientists to bring the discoveries of one field to bear upon those in another. A distressing example of the lack of coordination in our particular field was the reluctance of many scientists to grasp the significance of the discovery by Engelhardt and Ljubimova of the ATP-ase activity of myosin in 1939 [46] and later to reorientate their thinking in terms of the sliding filament mechanism, discovered by Hanson and Huxley in 1953 [56, 80]. This reluctance has been, and to some extent still is, a serious brake on further advance, although at last, after so many Quixotic tiltings at windmills, it now seems to have dawned on the doubters that the energy for contraction does indeed come from the splitting of ATP, when actin and myosin interact. Yet, the work of the schools of Szent Györgyi [143, 144, 145] and of Weber [160] from 1945 to 1953, had already demonstrated all the essential points, by the use of model systems of fibres and fibrils, where the structural elements were intact. Even the mechanism of relaxation had been worked out in these early studies, in the sense that it was proved that whenever the ATP-ase activity of myosin was inhibited, then ATP acted no longer as a contractile agent, but as a plasticizer or relaxant of the actin/myosin system [63, 155, 159, 160]. All the more important then that we should pay full tribute to these pioneers.

Bearing these remarks in mind, we can now set sail for a few of the uncharted regions, where gold and perhaps uranium lies buried. For this purpose, we shall divide the discussion into two parts, the first being a résumé of unsolved problems, and the second a discussion of thermodynamics and theories of contraction.

A. Unsolved problems

Sub-units of myosin. The question of whether or not myosin is made up of definite sub-units, in a similar manner to many other proteins, is of great importance in itself, and also throws a fascinating sidelight on genetic theory, because of the great length of the molecule; myosin is similar in this respect to the long molecules of tropocollagen. While it is certainly true that polysome systems are known which contain as many as seventy-five ribosomes [122], and that such systems are said to appear in developing muscle, particularly at the stage at which myosin filaments first become recognizable, it is not likely that they would be able to spin the whole molecule of 500,000 MW, and 1500 Å in length, in one go, head and tail combined. For one thing, we know the tail region probably consists of two polypeptide chains coiled about one another in helical form, and that these chains continue, in a less orderly fashion, right into the head region. The most to be expected of polysome systems would be to spin each of the strands of the helix in turn, which would then unite, and coil themselves round one another, as they peeled off the last ribosome in the series.

In common with many other examples, such as haemoglobin and myoglobin, it is probable that the individual ribosomes, involved here, would make polypeptide chains containing about 140 residues (= 16,000 MW), the amino-acid sequence of each sub-unit being unique [16, 120]. If that is so, each of the two or three strands in the myosin molecule could be built up by ten to fifteen ribosomes lying in a row, which would join the 16,000 MW sub-units together by peptide bonds. The last ribosomes in the series responsible for making the head region would need to be coded quite differently from the first eight or so, responsible for the tail, and it would also be amongst the 'head-region' ribosomes that varying codes could exist for the iso-myosins of varying specific ATP-ase activity [122].

Are any sub-units found which might correspond to those produced by this hypothetical genetic scheme? Indeed there are: dissociating agents for non-covalent bonds, particularly hydrogen-bond breakers such as 12 M urea or 5 M guanidine HCl, and even alkaline conditions of about pH 11, break up myosin into major sub-units of 160 to 260,000 particle weight, and also give rise to smaller units of particle weight between 16,000 and 22,000; the latter account for 10 to 20 per cent of the total weight [122, 172]. However, there is so much disagreement between the various workers about the exact particle weights of the two sub-units that one is left with a wide range of possibilities of fitting them into the parent molecule, though these are limited to some extent, according to whether one prefers a two- or three-stranded helical model for myosin. As we have seen (figure 1.2), the two-stranded model agrees best with the hydrodynamic, ORD and EM evidence; indeed, the EM shows the molecule to be double headed, with the two heads free to swivel about independently of one another. This in itself further limits the possible positions which could be occupied by the sub-units, and certainly makes it highly improbable that the smaller one could join the two heads together, which would otherwise have been a very attractive possibility.*

If we take 20,000 as the particle weight of the smaller sub-unit (c.f. 172) and suppose that there are two of these per molecule of 500,000, the remainder of the molecule would consist of two long chains, each of about 220,000 particle weight; the smaller unit would make up 8 per cent of the total; all of these figures lie in the experimental range. One could then suppose that the major effect of hydrogen-bond breakers such as urea is to unwrap the coiled-coil of the tail regions of each chain (LMM region), and perhaps also the highly coiled up polypeptide chains in each of the heads (HMM S-1), to give two very long sub-units of the required particle weight of 220,000.

But where do we fit in the smaller sub-units? It has been suggested that they might be attached along the sides of the heads, and held there by hydrogen-bonds [122]; at first sight this is a clumsy hypothesis, but it has the following advantages over any other; (i) it explains why urea and similar reagents dissociate the small unit; (ii) it avoids the great difficulty attendant on placing this unit somewhere in the main chain as a link, for instance, between the heads and the tails; not only would such a positioning upset the known particle weights of the larger units, but it is also unlikely that the

* Recent research (172; also, Scopes, R. K. and Perry, I. F., in press) gives particle weights of 16,000 and 22,000 for the active components.

longitudinally disposed hydrogen-bonds it would entail would be strong enough to bear the stresses suffered by the molecule during contraction; (iii) the side-by-side arrangement of the small units might serve to hold the coiled-up structure of the heads in place. For instance, if the particle weight of each head is taken as 120,000, as in figure 1.2, then its length fully extended would be 3300 Å; its measured length does not exceed 200 Å, so that it must be in a very highly coiled-up configuration in the native state as its high proline content would also suggest. A side-by-side link to a smaller sub-unit would clearly stabilize such an unusual structure. Note that HMM S-1 contains one proline 'kink' every 28 residues or so; hence the width of the molecule in its coiled up form would be about 84 Å [114].

End-groups and groups round the active site of myosin. The N-terminal sequence of both myosin and actin has been elucidated for a short stretch of the polypeptide chain. Both proteins exhibit the rather unusual feature that the N-terminal amino-acid residues, aspartyl in actin [1*a*], and seryl in myosin [118*a*], are acetylated; this accounts for the difficulty in the past of finding an N-terminal group in either of these proteins, and also disposes of the idea that the two polypeptide chains of the myosin backbone are linked together through N–N′ terminal linkages, which was muted at one time. It is not known at present whether the acetyl-N group of myosin is situated in the head or tail.

The sequence in myosin is acetyl-N-ser.ser.asp.ala.asp., and that in actin, acetyl-N-asp.glu-thr.ala. Both proteins have recently been shown to be unique in another respect, because they contain 3-methyl histidine [92*a*]. The latter amino-acid is very rare in nature, though it had earlier been shown to occur in mammalian blood. Actin contains 1 mole of it per mole, and myosin probably 2, 1 in each HMM S–1 fragment, i.e. 1 in each head.

Like the sequence studies, those on the groupings near the active site of myosin are still far from complete. The difficulty here is that very large molecules are often used to block off various groups; most of these reagents would be expected to affect the shape of the molecule near the active site, without necessarily being attached to it themselves, that is, they probably have an allosteric effect [122]. This is particularly likely in the case of SH-blocking reagents, such as *p*-chloromercuribenzoate which is known to have at first a slight activating effect on myosin ATP-ase and then later to inhibit it, as the concentration is raised and more SH-groups are blocked [122]. In this case, though, we have good evidence that two specific SH-groups (per half mole) are involved very near to the active site, because when $MgATP^{-2}$ is added to myosin it protects these two groups against attack by SH-reagents [142*a*, 147]. It seems quite possible that the ATP molecule is in fact attached to one of these groups through hydrogen bonding to the 6-NH_2 group of its purine ring, and to the other through a similar type of bond to one of the hydroxyl groups of the ribose moiety, the implication being that each of the heads of the molecule has a separate active site, able to split an ATP molecule.

By the use of a specific SH-reagent, bis *β*-carboxyl ether disulphide, it has also been shown that an extra SH-group is protected from attack, when actin is also present [142*a*]. This suggests that each of the heads of myosin has 2 of its SH-groups specifically oriented to bind to ATP, and another SH- near by, which binds to actin. There is other evidence of a similar nature which shows conclusively that the actin-binding site is different from the ATP-ase site [122]. In most cases, the quantitative results confirm that there is one of each of these sites per head (i.e. per particle weight of 250,000). This is a most important finding which is vital to the theoretical aspects of contraction.

Attempts have also been made to identify specific lysyl and histidyl residues near to the active site, the former by trinitrophenylation or guanidation [99, 122]. Such a drastic procedure as this would naturally tend to upset the whole structure of the head region, where there are a large number of lysyl residues; none of the inhibitory effects it produces can therefore be taken as conclusive evidence of this grouping at the active centre. The evidence for a histidyl residue is rather better [142]; indeed, in spite of the rather inconclusive nature of the actual findings, one might expect to find not only histidyl and lysyl residues at the active centre, but also arginyl, because all these basic groups could help to bind the negative charges of the triphosphate moieties of ATP^{-4} or $MgATP^{-2}$, whichever is the true substrate for actomyosin ATP-ase.

This particular aspect of ATP-ase activity is one of the fields where exciting new results may soon be expected; it is here where the very kernel of the problem of energy transduction is to be found.

Reaction mechanism. The mechanism of the splitting reaction depends so much on the nature of the active centre of myosin for its elucidation that it is premature to speculate too widely about it. Suggestions have frequently been made that the first step is the formation of a phosphoryl protein, as ATP splits to ADP and P_i, and that it is this phosphorylated intermediate which reacts with actin to produce movement [160, 150]. Most of the evidence for such an intermediate is derived from exchange studies with $H^{18}OH$ [98]; as we saw earlier, the ^{18}OH group of the labelled water appears in the inorganic phosphate released on splitting, and not in the ADP (equation 2.2). The essential question is, however, whether only this one labelled oxygen appears in the P_i or more than one; if the latter, then an intermediate phosphorylation of myosin must have occurred, with elimination of water. There is evidence that this happens during the very slow splitting of ATP by myosin in the presence of Mg ions, but not in the presence of Ca^{++} [98]. Other investigators can find no evidence for the formation of such a covalent phosphoryl bond [138].

There is other evidence for the formation of a phosphorylated intermediate during the splitting reaction from studies of the relatives rates of release of P_i and of the proton (cf. equation 2.2). Thus if the pH changes and the release of P_i are followed in a system where the splitting rate is not too fast, e.g. myosin in the presence of Mg^{++}, it is found that the release of the proton* always precedes that of the P_i [96, 149]. From this it is concluded that the P_i-myosin complex dissociates sufficiently slowly to have a detectable life. But is this not what might be expected from the conventional formulation of an enzyme reaction, where it is usually assumed that the product-enzyme complex will always have a definite lifetime (cf. equation 2.1)? For this reason, it does not seem that the somewhat

* See also 173 for elegant stop-flow measurements.

tardy release of P_i necessarily supports the idea that a phosphorylated intermediate is involved in the contractile process, in spite of the complex theory built up around the results [96, 149]. Moreover, such a phosphorylated intermediate should back react with labelled ADP to give labelled ATP, by analogy with oxidative phosphorylation in mitochondria [171]. In practice, no such back reaction can be detected, despite repeated attempts to demonstrate it [63].

Another question of major importance in relation to reaction mechanism is that of the role of the divalent cations, Mg^{++} and Ca^{++}, where as we have seen, Ca^{++} activates the splitting of ATP by myosin, either alone or in the presence of actin, but is not capable of initiating movements of the heads and sliding of the two sorts of filament past one another, unless Mg^{++} is also present. Apart from the suggestions made earlier (chapter 2) [31*a*] there is at present no satisfactory explanation of these confusing phenomena, although recent investigations provide some possible answers [149, 168]. It is certainly a problem which needs tackling urgently. However, one essential point about the binding process can now at last be cleared up: that is the question of the structure of the Ca- and Mg-chelates of ATP. These have been shown by nuclear magnetic resonance studies [28*a*] to consist of chelates between the OH-groups of the two terminal P-atoms of the triphosphate; neither of them involve the NH_2-group of the purine ring.

B. *Thermodynamics and theories of contraction*

Although we are still largely ignorant of the exact dimensions of the three important parameters, the changes in internal energy ($-\Delta E$), in free energy ($-\Delta F$) and in entropy (ΔS), which arise from the underlying chemical reactions, it is nevertheless useful to consider the thermodynamic basis of the contractile process, since this fixes the framework within which these reactions must take place.

We have discussed the first law in chapters 7 and 8, and shown how it applies to the energetics of contraction. Briefly, the energy given out in a contraction is made up of two terms, the heat ($-Q$) and the actual work done ($+W$); the relation of these two parameters to one another is all that we can actually observe and measure, while the contraction is going on. Both terms depend on the underlying chemical reactions: the splitting of ATP, which yields the energy for contraction and the resynthetic Lohmann reaction which quickly follows in its wake. As in the case of any other reversible chemical reaction, the maximal work the splitting reaction can be made to perform, through transduction of energy to the contractile elements, is fixed by the release of *free energy* ($-\Delta F$), as the reaction precedes.

Free energy. The free energy, $-\Delta F$, released during the course of a chemical reaction, may be derived from the van't Hoff isotherm, if the equilibrium constant of the reaction, K, is known. (Note that any free energy, not 'transduced' into work, will appear as heat.)

The necessary terms for the Lohmann reaction, in the case where ATP is being reformed from ADP and PC, can be formulated† as follows:

$$\left.\begin{array}{ll}(a) & K = [\mathrm{ATP}][\mathrm{creatine}]/[\mathrm{ADP}][\mathrm{PC}] \quad \text{(at equilibrium)} \\ (b) & -\Delta F^\circ = RT \ln K \\ (c) & -\Delta F = -\Delta F^\circ - RT \ln\, [\mathrm{ATP}][\mathrm{creatine}]/[\mathrm{ADP}][\mathrm{PC}] \\ & \qquad \text{(at any stage of reaction)}\end{array}\right\} \quad 9.1$$

Equation (*c*) shows us that the actual free energy obtained per mole on going from a particular stage to the equilibrium state is dependent on a constant term, the standard free energy change, $-\Delta F^\circ$, and on another term which depends on the relative concentrations of products and reactants at that stage. When the reaction has reached equilibrium, it is clear that the second term must, by definition, be equal in magnitude, but opposite in sign, to $RT \ln K$ ($= -\Delta F^\circ$); hence there can be no further change in free energy ($-\Delta F$), and the reaction stops. At any other stage the second log term may be negative (ratio of products to reactants less than 1·0), or positive (ratio greater than 1·0); this will either add to or subtract from the actual release of free energy, in amounts depending on the magnitude of $-\Delta F^\circ$, itself dependent on the equilibrium constant.

We can think of the free energy changes in another way by con-

† Simplified form; the full equation should include terms for MgATP^{-2} and MgADP^{-1}.

sidering the free energy, due to the second term on the right-hand side of equation 9.1*c*, as representing the work done to take the reaction from any particular stage to equilibrium. This, therefore, uses up part of the standard free energy change ($-\Delta F^\circ$), the remainder being available for external work.

Under physiological conditions, K for the Lohmann reaction is about 100, giving $-\Delta F^\circ = 2{\cdot}5$ kcal/mole at 0°C. In the resting state, a frog sartorius muscle has a concentration of ATP of about 3 mM, of creatine of 10 mM and of PC of 20 mM; from the given equilibrium constant, this means that the free [ADP] in the muscle cannot exceed $1{\cdot}5 \times 10^{-2}$ mM.

The equilibrium state is, of course, immediately upset by a single twitch or short tetanus, during which ATP is split and its concentration falls to about 2·5 mM, while that of ADP rises to about 0·5 mM. This will set the Lohmann reaction going towards the right, but how much free energy will be released will depend on whether the reaction starts as soon as ADP is produced in excess of the equilibrium concentration of $1{\cdot}5 \times 10^{-2}$ mM, or whether there is a considerable delay, during which [ADP] can build up (as pH measurements on contracting muscle suggest [32, 91]). In the first case, the reaction is operating near to equilibrium, so that less free energy will be liberated per mole than in the case of a delay, as equation 9.1*c* shows. In either event, it is unlikely that the free energy released can be transduced into work, but is degraded into heat. Even though the quantities may be small (the maximum is about 2 kcal/mole in the case of a long delay), it is not justifiable to assume, as many treatments of the subject do, that the Lohmann reaction is thermally neutral.

The splitting of ATP, which is the immediate source of the free energy which is transduced into work during contraction, differs fundamentally from the Lohmann reaction, because it is hydrolytic (equation 8.1*a*). The concentration terms to be inserted into the van't Hoff isotherm become $[ADP][P_i]/[ATP][H_2O]$, and since the first three are very low (3 mM or less) and the concentration of H_2O in muscle is high (at least 44 M), it follows that the equilibrium of the splitting reaction is far to the right in favour of the products. In fact, the reaction for most of its course is pseudo-unimolecular, so that the log term on the right-hand side of equation 9.1*c* is large, negative and almost constant; hence the value of $-\Delta F$ will be larger than $-\Delta F^\circ$ by a considerable margin. Unfortunately we do not know the value of $-\Delta F$ at all accurately; uncorrected *in vitro* values range between 7 and 10·5 kcals/mole [125, 169].

As we shall see, the uncertainty in $-\Delta F$ for the splitting of ATP leaves us in doubt about the magnitude of the entropy changes which may occur during its course. The only reliable guide we have is the total energy, or enthalpy, change ($-\Delta E$) in living muscle for the splitting and Lohmann reactions combined, which is 10·5 kcal/mole [25]. The maximal contribution of the Lohmann reaction to this value is about 2 kcal/mole, leaving 8·5 kcal/mole for $-\Delta E$ for the splitting reaction alone; this falls within the above range of $-\Delta F$ values, and suggests that the entropy change during splitting is either zero or very small.

Other considerations which need to be taken into account for a quantitative appraisal of the free energy changes are the amounts of heat liberated due to neutralization of the protons released during the splitting and Lohmann reactions. The combined effect is the absorption of 0 to 0·2 protons per mole split, but in FDNB-poisoned muscles, where the Lohmann reaction is inhibited, 0·8 to 1·0 protons are released during splitting, and neutralization of these may make a significant contribution to the total energy changes [125]. On the other hand, the heat from this source may, in fact, just balance the missing heat output from the Lohmann reaction (maximum 2 kcal/mole), thus giving nearly identical values for $-\Delta E$ in the two cases.

Activation energy or enthalpy. It was assumed in the previous section that a reaction would proceed so long as the free energy change ($-\Delta F$) on passing from reactants to products, was positive. This is an over-simplification, except in the case of a few spontaneous reactions, where the so-called activation energy or enthalpy is very low: an example is the 'spontaneous' combustion of white phosphorus when placed in air. In most cases, however, molecules will not react until they have acquired extra energy over and above that which they possess in the ground state: this extra energy is the *energy or enthalpy of activation.* Often it is supplied, in the examples familiar in a chemical laboratory, by heating up the reactants, and thus increasing the rate of collision between them, through which they acquire the necessary extra energy to react. Such a crude procedure does not occur in living systems; instead, use is made of the specific catalysts we know as enzymes. The mode of action of all catalysts is to bring the reacting molecules closer together, that is, to increase their order and decrease their configurational entropy. As a result of this new order, new stresses are set up, which in the case of hydrolytic reactions, such as the splitting of ATP, destabilize the substrate molecule, and thereby drastically reduce the extra energy needed for activation.

For the splitting of ATP, the activation energy, when the reaction is catalysed by actomyosin or fibrils, is about 20 kcal/mole; in the absence of enzyme, on the other hand, it would be of astronomical proportions, because we know that ATP can be boiled in dilute buffers for many days at physiological pH without splitting to any significant extent. Similar considerations apply to the Lohmann reaction, and also to most of the enzyme-catalysed reactions of the glycolytic cycle. Note that the free energy change of reaction ($-\Delta F$) remains unaltered, whether enzyme is present or not.

Entropy and the second and third laws. We must now pass into the more hazardous realm of the second and third laws. In its bald form, the second law states that all systems tend towards the greatest state of disorder; the measure of this disorder which accompanies a chemical or physical change is the entropy change, ΔS. A positive value of ΔS means that the system has become more disordered as a result of the change, whereas a negative value shows it has become more ordered or 'crystalline'. A good example of a positive entropy change is the melting of a crystal, which obviously becomes more disordered in the molten state. The reverse process of crystallization is a negative entropy change, that is a *decrease in the disorder* (=an increase in order) of the system.

In the case of the enzymes and proteins in which we are interested, the orderly 'crystal lattice' of the native molecule is held together mainly by hydrogen bonding within and between peptide chains, and is further stabilized by ionic interactions between oppositely charged side-chain groups, and also by van der Waals forces between non-polar regions. When the temperature is raised sufficiently, usually in the region of 60°C, the increased heat motion and vibrational energy of the molecule suddenly causes a number of hydrogen bonds to break so that the highly complex secondary and tertiary structure of the molecule tends to collapse into a more disorderly, random form. A good example is the melting of the long, orderly triple helix of collagen; with the collagen fibres from mammals this sets in at about 61°C, and results in very rapid shrinkage; during the process the fibre can develop as much as 10 kg of force per cm^2 under isometric constraint, and can do a great deal of work under isotonic conditions. The shrinkage represents a positive entropy change, whereas the work is done at the expense of the free energy released ($-\Delta F$ is positive).

The relation between the entropy changes and those in free and internal energy must be derived here intuitively. We invoke the help of the third law, or Nernst heat theorem, which states that systems tend to the greatest state of *order*, that is to their *lowest* entropy, as the temperature is reduced towards absolute zero: at this temperature the entropy and internal energy of ideal systems becomes zero; in other words, they become maximally ordered. The corollary is that the entropy of a system, and with it the internal energy, increase as the temperature is raised. Hence, one might guess that multiplication of the entropy change (ΔS) by the absolute temperature (T) would yield a product, $T\Delta S$, with the dimensions of energy. Indeed this is so; $T\Delta S$ is a heat term, and therefore, whenever the entropy of a system changes, heat is given out or taken in to this extent, and the internal energy (E) falls or rises correspondingly, depending on whether the term is positive or negative.

We can formulate the change in internal energy when a reaction takes place, as follows:

$$
\begin{aligned}
-\Delta E &= -\Delta F - T\Delta S \qquad &9.2\\
&= -Q + W \qquad &\text{(from first law, equation 7.1)}
\end{aligned}
$$

As we said, $-\Delta F$ is the maximal work which can be obtained from a reaction, or in other words, it is the limiting value of the work (W) which can actually be done. It is rare to find systems where such a state of affairs, a thermodynamic efficiency ($-W/\Delta F$) of 1, can be realized, so in most real systems some of the released free energy will be degraded into heat, with the result that the mechanical efficiency ($-W/\Delta E$) will always be less than the thermodynamic efficiency, even when the $T\Delta S$ term in equation 9.2 is zero.

To apply the above considerations to the contractile system, we need to know at least two of the parameters in equation 9.2, for the special cases of the splitting and Lohmann reactions. At present it cannot be said that we possess sufficiently accurate data to enable us to make more than a guess at the true state of affairs. For instance, taking $-\Delta F$ for the splitting reaction as 7 to 10·5 kcal/mole [169], and that for the Lohmann reaction as about 2 kcal/mole, the free energy change for both combined would lie between 9 and 12·5 kcal/mole. Since the observed total enthalpy change is 10·5 kcal/mole, equation 9.2 would give values between $-1{\cdot}5$ and $+2{\cdot}0$ kcal/mole for $T\Delta S$, representing changes of between -5 and $+7$ entropy units at 0°C. This makes it unlikely that the entropy changes differ significantly from zero; that is to say, $-\Delta E$ for the two reactions is nearly the same as $-\Delta F$. The implication of this is that the heat output during contraction must come from degraded free energy and not from 'degraded' entropy.

Theories of contraction. Of all the phenomena which a theory of contraction must take into account, the most important is the transduction process and with it, the movements of myosin heads which are directly responsible for pulling or pushing the actin monomers towards the centre of the A-band. At present, there is only one theory which attempts this task in anything but a perfunctory manner [31], although it had a brilliant precursor which well repays eloser study [85*a*].

The theory proposes, as its major premise, that there is a short, rather disorderly section of polypeptide chain in the region of the active site of myosin, which can adopt two quite distinctive configurations (figure 9.1). In the absence of Ca ions, in the resting state, this polypeptide is kept in its random, but extended and 'snaking' form, because the negative charge of $MgATP^{-2}$, attached

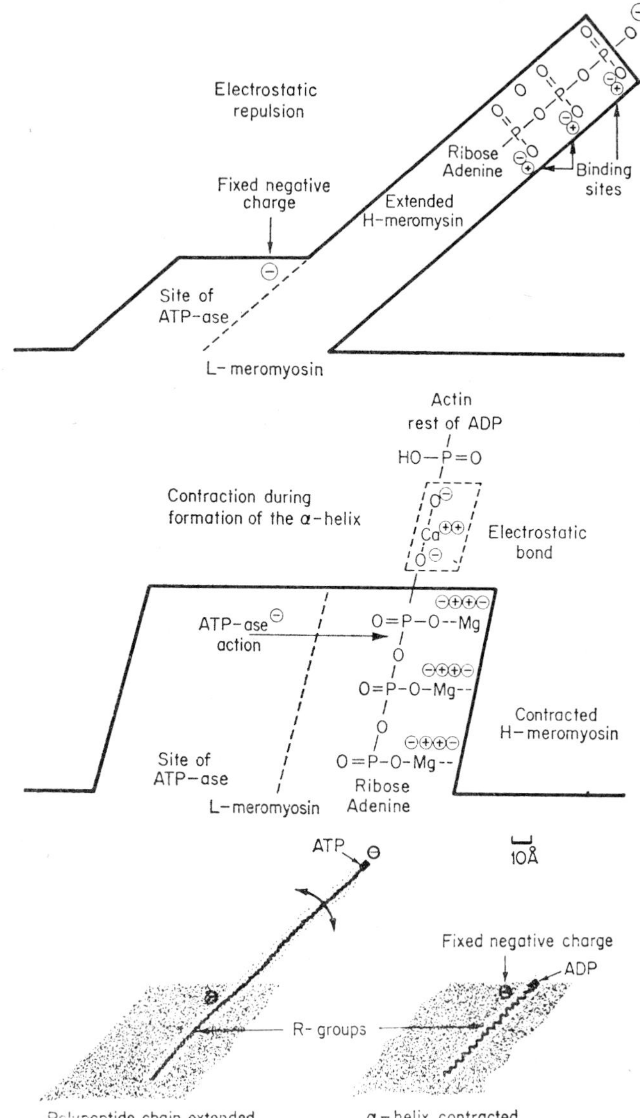

Figure 9.1 (*a*) Postulated resting, extended state of heads of myosin molecules, with ATP bound to the reacting site.

(*b*) Postulated contracted state of heads of myosin molecules, after formation of Ca-link with actin-ADP, and contraction to the α-helical form.

(*c*) Another way of envisaging the terminal polypeptide chain of a myosin head, in its extended, random form, and in the contracted α-helical form. (After [31].)

at the active site, is repelled by another negative, 'fixed' charge, somewhere in the neighbourhood. When activation begins, a Ca ion arrives, attaches itself to the bound ATP and forms a link between it and the bound ADP of actin. This neutralization of the negative charge on $MgATP^{-2}$ means that the random polypeptide can now contract and partly reform itself into an α-helix, by hydrogen bond formation; this releases the potential energy, stored in the extended polypeptide, as a result of the original change repulsion (figure 9.1*a*). The shortening of the polypeptide by α-helix formation has, first, the effect of pulling the actin along one step, in relation to the myosin heads, and secondly, of dragging the attached ATP into the region of a hypothetical ATP-ase site, in a near-by portion of the head, and to one side of it. Note that by correct positioning of this site and also of the slope of the heads in relation to the rest of the myosin filament, the effect of the collapse to a helical form is to pull the attached actin towards the M-line.

The final step in this part of the cycle is the splitting of ATP at the ATP-ase site, and its rapid replenishment by a fresh molecule of $MgATP^{-2}$. The cycle will repeat itself, while Ca ions are still present, but when they are withdrawn during relaxation, the attached $MgATP^{-2}$ is once more repelled by the fixed negative charge, the α-helix is broken, and the newly established random polypeptide can snake out again into its extended form. The process is shown in diagrammatic form in figure 9.2, which is self-explanatory.

Calculations are given in detail by the author to show that the implied changes in free energy and entropy are possible on the basis of what is already known about the formation of α-helices in synthetic polypeptides. How the free energy of ATP is actually released and used is not so clearly explained. Moreover, the energy release during the ATP-ase step seems to have been overlooked, unless one supposes that this is so small, because of the earlier distortion of ATP during the charge repulsion step, that it can be neglected. However, it is not our purpose here to be drawn into arguments about the quantitative energy changes during a hypothetical scheme of this sort, because it seems obvious that if we do not know with certainty even the elementary facts about the free

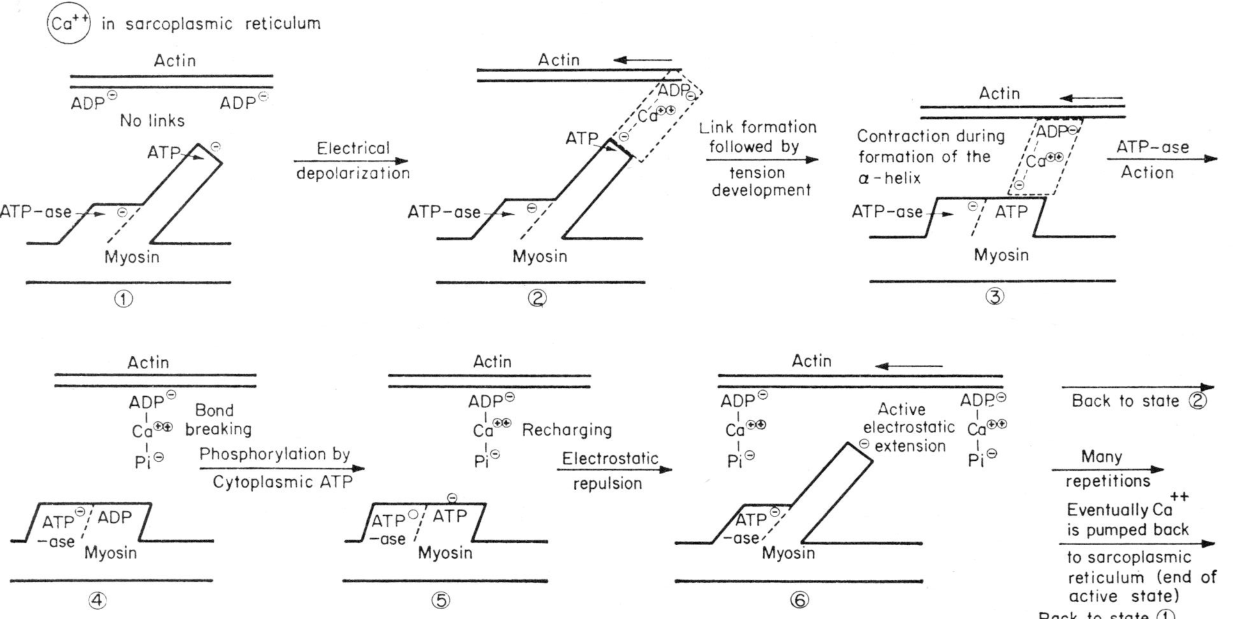

Figure 9.2. The stepwise series of reactions which would lead to a single miniature contraction cycle of one of the myosin heads of a filament, pulling the attached actin along a distance of about 100 Å. The recovery process is also depicted. After [31].

energy and entropy changes during the splitting of ATP by a simple system such as myosin, we cannot possibly be in a position even to calculate approximately what might happen to these parameters during the changes suggested by figure 9.1.

More specific criticisms of the proposed reaction mechanism are that it fails to take into account the separate actin-combining and ATP-ase sites on myosin, and also to explain how the Ca-link can be formed when Mg^{++} is already attached to ATP^{-4}. It would not be difficult, however, to modify the theory in the second respect, in the sense we suggested earlier: that is, that the role of Ca^{++}, when tropomyosin and troponin are present in the actin helices, is to neutralize hypothetical negative charges on this 'troponin' and thus allow the close approach of the myosin heads, with $MgATP^{-2}$ attached, to the monomers of the actin super-helix. The question of the separate sites on the myosin heads is a more serious one.

The following alternative seems more feasible: (i) actin cannot become attached to its specific binding site on myosin while $MgATP^{-2}$ or ATP^{-4} is sitting on the near-by ATP-ase site, because of the charge repulsion we suggested; (ii) as soon as Ca^{++} has neutralized the charge on troponin, actin can attach itself to myosin; this attachment necessarily distorts the myosin head, and it is this which triggers off splitting at the ATP-ase site (see later); (iii) to make the system move, some kind of change from random-to-helical, or vice versa, would be necessary in the head region, but brought about, we suggest, by the free energy of splitting; (iv) unless fresh ATP appeared, the system would now be in a rigor state, with actin still attached to myosin at its binding site; as soon as $MgATP^{-2}$ was replenished, however, the cycle could merely repeat itself, with the Ca^{++} still in position on tropomyosin and neutralizing its charge. Such a modification could easily be accommodated within the main premises of the theory.

Attractive as this theory is, there are alternative ways in which the transduction process might take place. As we have seen, a cornerstone of the theory is a change from randomly coiled to α-helical chain in the head region of myosin: it is this which pulls the actin monomers along one step during each miniature contraction cycle. In other words, the process is conceived as taking place by a gain in order, that is, a loss of configurational entropy which necessarily entails an output of heat, because $-T\Delta S$ is positive: this is reversed and heat reabsorbed in the next step in the cycle and so on. The result is that there can be no net output of heat due to the entropy changes, because they are reversible in the above sense. So far so good, but there seems no reason why we should not take a diametrically opposed view of what goes on, and replace the initial negative entropy change with a positive one, by analogy with other protein systems.

Heat shrinkage of collagen is a good example of a system which can develop tension and do work as the result of the *collapse* of a helix, that is, through a gain in disorder and an increase in entropy. During the collapse, on heat shrinkage at about 61°C, the bonds which hold the triple helix of this protein together are broken by the increased heat motion and vibrational energy, with the result that non-polar groups previously buried within the helices are exposed, and tend to aggregate together, rather as oil does in water. In the case of beef tendon collagen, which is highly cross-linked, the molecule shrinks with great force, and develops a tension of 10 kg/cm^2 or more, or if allowed to contract isotonically, does large amounts of work. Looked at thermodynamically, the molecules have taken in heat energy during the entropy change ($-T\Delta S$), and their internal energy (E) has, thereby, temporarily increased; as they shrink and do work or develop tension, some of this newly acquired internal energy is lost again as free energy ($-\Delta F$), available for external work. In other words, the system works as an entropy machine. The change is partly reversible, because when the temperature is lowered, the internal energy falls, and this allows some spontaneous re-formation of the triple helix.

Applying this analogy to muscle, we propose that the first change during each miniature contraction cycle is the collapse of a helical portion of polypeptide chain, in the head region of myosin, to a more random, but highly coiled configuration. This would take place after charge neutralization by Ca^{++}, whenever a myosin head became temporarily attached to an actin monomer, through (*a*) a non-covalent bond to its actin combining centre, and (*b*) through the binding of the $MgATP^{+2}$ or ATP^{-4}, already attached at its myosin site, to a centre on actin possibly through an ionic bond. It must then be assumed that it is the latter bonding which causes destabilization of the helix; this then collapses in the same way as heat-shrinking collagen, and during the process pulls the ADP moiety of ATP off with it, leaving the terminal phosphate group behind on the actin partner. The free energy released is either degraded into heat, most of which is reabsorbed by the contracting helical portion of the head, as it goes through its positive entropy change, or else is absorbed directly, as the internal energy (E) of the polypeptide chain increases. Part or all of the increased internal energy is then released again as free energy ($-\Delta F$) during the collapse, and this is used for external work; any unused free energy will be degraded into heat.

At the end of the process of collapse, the internal energy of the shrunken polypeptide will have fallen, through the release of free energy, to nearly the same level as that of the original helix, so that re-formation of the latter can occur almost spontaneously, without large energy changes. This results in re-extension of the head, as it reacquires $MgATP^{-2}$ or ATP^{-4} from the sarcoplasm, ready to go through another cycle.

The advantage of this scheme is first, that it is analogous to a known process and requires the minimal number of steps, whereas the formation of helix from random chain during the energy yielding part of the contraction cycle, as proposed in the theory we have discussed, is something quite new, and involves several highly controversial energy changes; secondly, the scheme suggests a simple way in which the free energy of ATP-splitting can be used to do work, through the operation of an entropy machine, in contrast to the negative entropy change, proposed in the other theory, which must necessarily result in the wastage of energy as heat, because $-T\Delta S$ is positive. Note that both theories apply strictly only to the overall cycle; they do not take shortening heat into account, in spite of assertions to the contrary, because in both, the entropy changes from which this might arise are instantaneously reversible.

Whichever scheme is adopted, suppositions have to be made about the possible number of links formed between actin and myosin at different velocities of shortening, themselves dependent on the load. If, for instance, the degree of overlap of actin and myosin is read off from figure 6.7, and compared with the tension developed, it is seen that for the muscle to develop a tension equal to that in a small load and thus to begin to lift it, only a small degree of overlap is necessary, whereas for it to develop the maximal isometric tension, maximal overlap is required: since overlap is a direct reflection of the potential formation of links between myosin heads and actin monomers, it is clear that at low loads and high velocities of shortening only very few links are formed at any instant during shortening, and this in its turn limits how much ATP can be split and how many miniature cycles of entropy change can take place. At very high loads on the other hand, the potential formation of links approaches the maximal observed under isometric conditions, but now the heavy load so limits the velocity and extent of shortening that it must automatically slow down the rate of link breakage during the final phase of each miniature cycle, and thus once more limit the ATP breakdown on which this breakage depends.

As we have seen in chapters 7 and 8, this is exactly what happens in practice; the total energy given out, and the total ATP split, during a twitch at either extremes of loading, are both much lower than at intermediate loads, where the maximal work can also be done. The latter effect comes about because, at these intermediate loadings, the two opposing effects of increasing potential linkage formation and decreasing possibility of breakage just balance each other. These suppositions are a direct outcome of the sliding filament theory, and not of either of the theories of energy transduction discussed here, but it is at least encouraging that they can both accommodate them. It remains to be seen whether experiment can decide between the two opposite types of entropy change proposed, and thus set the stage for the next step forward.

If we have unearned luck
Now to 'scape the serpent's tongue,
We will make amends ere long;
Else the Puck a liar call:
So, good night unto you all.

W. S.

Bibliography

1. ABBOTT, B. C., AUBERT, X. and HILL, A. V., 1951. 'The absorption of work by a muscle stretched during contraction': *Proc. R. Soc. B.* 139, 86.

1a. ALVING, R. E. and LAKI, K., 1966. 'The N-terminal sequence of actin': *Biochem.* 5, 2597.

2. AUBERT, X., 1956. 'Le mécanisme contractile *in vivo*: aspects mécaniques et thermiques': *J. Physiol (Paris)* 48, 105.

3. AUBERT, X., 1956. 'Le couplage énergétique de la contraction musculaire': Thèse d'agrégation, Université Catholique de Louvain. Bruxelles. Editions Arscia.

4. AUBERT, X., 1964. 'Tension and heat production of frog muscle tetanized after intoxication with FDNB': *Pflügers Arch. ges. Physiol.* 281, 13.

5. BÁRÁNY, M., NAGY, B., FINKELMANN, F. and CHRAMBACH, A., 1961. 'Studies on the removal of the bound nucleotide of actin': *J. biol. Chem.* 236, 2917. See also: BÁRÁNY, M., TUCCI, A. F., and CONOVER, T. E., 1966. *J. molec. biol.* 19, 493.

6. BÁRÁNY, M., GAETJENS, E., BÁRÁNY, K., and KARP, E., 1963. 'Comparative studies of rabbit cardiac and skeletal myosins': *Arch. Biochem. Biophys.* 106, 280.

7. BÁRÁNY, M., FINKELMANN, F. and THERATTIL-ANTHONY, T., 1962. 'Studies on the bound calcium of actin': *Arch. Biochem. Biophys.* 98, 28.

8. BARANYI, E. H., EDMAN, K. A. P. and PALIS, A., 1951. 'The effect of ATP and metal ions on viscosity of solutions of actomyosin': quoted by D. M. Needham in *Structure and Function of Muscle*, vol. II, ed. G. H. Bourne, Academic Press – New York (1960), pp. 67 and 96.

9. BAKER, P. F., HODGKIN, A. L. and SHAW, T. I., 1961. 'Replacement of the protoplasm of a giant nerve fibre with artificial solutions': *Nature Lond.* 190, 885.

10. BENDALL, J. R., 1961. 'A study of the kinetics of the fibrillar ATP-ase of rabbit skeletal muscle': *Biochem. J.* 81, 520.

11. BENDALL, J. R., 1964. 'The myofibrillar ATP-ase activity of various animals in relation to ionic strength and temperature': in *Biochemistry of*

Muscular Contraction: ed. J. Gergely. Little Brown & Co. – Boston, Mass., p. 87.

12. BENDALL, J. R., 1958. 'Relaxation of glycerol-treated muscle fibres by ethylenediamine tetraacetate': *Arch. Biochem. Biophys.* 73, 283.
13. BENDALL, J. R., 1953. 'Further observations on a factor effecting relaxation of ATP-shortened muscle-fibre models and the effect of Mg and Ca ions upon it': *J. Physiol.* 121, 232.
14. BENDALL, J. R., 1964. 'Post-mortem changes in muscle': in *Structure and Function of Muscle*, vol. III, p. 227.
15. BIRKS, R., HUXLEY, H. E. and KATZ, B., 1960. 'The fine structure of the neuromuscular junction of the frog': *J. Physiol.* 150, 134.
16. BISHOP, J., LEAHY, J. and SCHWEET, R., 1960. 'Formation of the peptide chain of haemoglobin': *Proc. Nat. Acad. Sci. U.S.A.* 46, 1030.
17. BOWEN, W. J., 1964. 'Glycerol-treated muscle as a model of contraction': in *Biochemistry of Muscular Contraction*, p. 441.
18. BOYER, P. D., 1964. 'Oxygen exchange and oxidative phosphorylation studies as related to possible phosphorylations accompanying muscle contraction': *Biochemistry of Muscle Contraction*, ed. J. Gergely, Little Brown & Co. – Boston, Mass., p. 94.
19. BOZLER, E., 1953. 'Evidence for an ATP-actomyosin complex in relaxed muscle and its response to Ca ions': *Am. J. Physiol.* 168, 760.
20. BRIGGS, F. N. and FUCHS, F., 1967. 'The site of calcium-binding in the activation of myofibrillar contraction': *Fed. Proc.* 26, 598.
21. BRIGGS, F. N. and PORTZEHL, H., 1957. 'The influence of relaxing factor on the pH dependence of the contraction of muscle models': *Biochim. biophys. Acta* 24, 482.
22. BURGE, R. E. and ELLIOT, G. F., 1963. 'The thermal dependence of (i) the decay of the twitch and tetanic tension, and of (ii) elongation after stimulation': *J. Physiol.* 169, 86B.
23. CAIN, D. F., INFANTE, A. A. and DAVIES, R. E., 1962. 'Chemistry of muscle contraction': *Nature Lond.* 196, 214.
24. CARLSON, F. D., HARDY, D. and WILKIE, D. R., 1963. 'Total energy production and PC hydrolysis in the isotonic twitch': *J. gen. Physiol.* 46, 851. See also, *J. Physiol.* 195, 157 (1968).
25. CARLSON, F. D., HARDY, D. and WILKIE, D. R., 1967. 'Heat production and PC splitting in muscle': *J. Physiol.* 189, 209.
26. CARLSON, F. D. and SIGER, A., 1960. 'The mechanochemistry of muscular contraction. I. The isometric twitch': *J. gen. Physiol.* 44, 33.
27. COUTEAUX, R., 1966. 'The structure of the motor end-plate': in *Symposium on Muscle*, 1966 (Hungarian Acad. of Sciences, Budapest – in press).
28. COUTEAUX, R., 1960. 'Motor end-plate structure': in *Structure and Function of Muscle*, vol. I, 337–78.

28*a*. COHN, M. and HUGHES, T. R. (jr.), 1962. 'Nuclear magnetic resonance spectra of ADP and ATP. II. Effect of complexing with divalent metal ions': *J. biol. Chem.* 237, 176.

29. DAVIES, R. E., 1965. 'The role of ATP in muscle contraction': in '*Muscle*', *Symposium held at University of Alberta*, Pergamon Press – London, pp. 49–69.

30. DAVIES, R. E., KUSHMERICK, M. J. and LARSON, R. E., 1967. 'ATP, activation and the heat of shortening of muscle': *Nature Lond.* 214, 148.

31. DAVIES, R. E., 1963. 'A molecular theory of muscle contraction': *Nature Lond.* 199, 1068.

32. DISTÈCHE, A., 1960. 'Contribution à l'étude des échanges d'ions hydrogène au cours du cycle de la contraction musculaire': Memoires de l'Académie Royale de Belgique. Tome XXXII.

33. DRABIKOWSKI, W. and GERGELY, J., 1964. 'The effect of temperature of extraction and of tropomyosin on the viscosity of actin': in *Biochemistry of Muscle Contraction*, p. 125.

34. DYDYNSKA, M. and WILKIE, D. R., 1966. 'The chemical and energetic properties of muscle poisoned with FDNB': *J. Physiol.* 184, 751.

35. EBASHI, S. and KODAMA, A., 1966. 'Native tropomyosin-like action of troponin on trypsin-treated myosin B (actomyosin)': *Jap. J. Biochem.* 60, 733.

35*a*. EBASHI, S., EBASHI, F. and KODAMA, A., 1967. 'Troponin as the Ca receptive protein in the contractile system': *Jap. J. Biochem.* 62, 137. See also *Biochem. biophys. res. comm.* 31, 647.

36. EBASHI, S., 1960. 'Calcium binding and relaxation in the actomyosin system': *Jap. J. Biochem.* 48, 150.

37. EBASHI, S., 1961. 'Relaxation and removal of calcium by muscle particulate fraction': *Jap. J. Biochem.* 50, 77.

38. EBASHI, S., 1963. 'Third component participating in the superprecipitation of natural actomyosin': *Nature Lond.* 200, 1010.

39. EBASHI, F. and EBASHI, S., 1962. 'Removal of calcium and relaxation in actomyosin systems': *Nature Lond.* 194, 378.

40. EBASHI, S. and LIPMANN, F., 1962. 'ATP-linked concentration of Ca ions in a particulate fraction of rabbit muscle': *J. Cell. Biol.* 14, 389.

41. ECCLES, J. C., 1952. 'The ionic hypothesis and the active membrane': *The neurological basis of mind*, Clarendon Press – Oxford, p. 35.

42. EDMAN, K. A. P., 1966. 'The relation between sarcomere length and active tension in isolated semitendinosus fibres of the frog': *J. Physiol.* 183, 407.

43. ELLIOTT, G. F., LOWY, J. and WORTHINGTON, C. R., 1963. 'X-ray diffraction studies of resting, living muscle': *J. molec. biol.* 6, 295.

44. ELLIOTT, G. F., LOWY, J. and MILLMANN, B. M., 1967. 'Low angle X-ray diffraction studies of living striated muscle during contraction': *J. molec. biol.* 25, 31.

45. ENDO, M., NONOMURA, Y., MASAKI, T., OHTSUKI, I. and EBASHI, S., 1966. 'Localisation of native tropomyosin in relation to striation patterns': *Jap. J. Biochem.* 60, 605.

46. ENGELHARDT, W. A. and LJUBIMOVA, M. N., 1939. 'Myosin and adenosine triphosphatase': *Nature Lond.* 144, 669.

47. FENN, W. O., 1924. 'The relation between the work performed and the energy liberated in muscle contraction': *J. Physiol.* 58, 373.

48. FRANZINI-ARMSTRONG, C. and PORTER, K. R., 1964. 'The Z-disc of skeletal muscle fibres': *Zeit. für Zellforsch.* 61, 661.

49. FRANZINI-ARMSTRONG, C. and PORTER, K. R., 1964. 'Sarcolemmal invaginations and the T-system in fish skeletal muscle': *Nature Lond.* 202, 355.

50. GIBBS, C. L., RICCHINTI, N. V. and MOMMAERTS, W. F. H. M., 1966. 'Activation heat in frog sartorius muscle': *J. gen. Physiol.* 49, 517.

51. GLASSTONE, S., 1947. 'Theory of absolute reaction rates': *Textbook of Physical Chemistry*, 2nd edition, D. Van Nostrand & Co. Inc., New York, p. 1098 et seq.

52. GORDON, A. M., HUXLEY, A. F. and JULIAN, F. J., 1966. 'The variation in isometric tension with sarcomere length in vertebrate muscle fibres': *J. Physiol.* 184, 170.

53. GREEN, I. and MOMMAERTS, W. F. H. M., 1953. 'Release of hydrogen ions during the splitting of ATP': *J. biol. Chem.* 202, 541.

54. GUBA, F. and HARSANYI, V., 1966. 'Myofibrillin – a new structural protein': in *Symposium on Muscle*, 1966 (Hungarian Acad. of Sci., Budapest – in press).

55. HANSON, J. and LOWY, J., 1965. 'Electron microscope studies of bacterial flagellae': *J. molec. biol.* 11, 293.

56. HANSON, J. and HUXLEY, H. E., 1955. 'Sliding filament mechanism of muscle contraction': *Symp. Soc. exp. Biol.* 9, 228.

57. HANSON, J. and LOWY, J., 1965. 'Molecular basis of contractility in muscle': *Br. med. Bull.* 21, 264.

58. HANSON, J. and HUXLEY, H. E., 1960. 'The molecular basis of contraction in cross-striated muscles': *Structure and Function of Muscle*, vol. I, pp. 183–225.

59. HANSON, J., 1967. 'Axial period of actin filaments': *Nature Lond.* 213, 353.

60. HASSELBACH, W., 1952. 'Die Diffusionskonstante des Adenosintriphosphats im Innerer der Muskelfaser': *Z. Naturf.* 7B, 334.

61. HASSELBACH, W., 1956. 'Die Wechselwirkung verscheidener Nukleosidtriphosphate mit Actomyosin im Gelzustand': *Biochim. biophys. Acta* 20, 355.

62. HASSELBACH, W. and MAKINOSE, M., 1961. 'Die Calciumpumpe der "Erschlaffungsgrana" des Muskels und ihre Abhängigkeit von der ATP-spaltung': *Biochem. Z.* 333, 518.

63. HASSELBACH, W., 1964. 'Relaxing factor and the relaxation of muscle': *Prog. Biophys. molec. Biol.* 14, 169.

64. HASSELBACH, W. and SERAYDARIAN, K., 1966. 'The rôle of sulphydryl groups in calcium transport through the sarcoplasmic membranes of skeletal muscle': *Biochem. Z.* 345, 159.

65. HASSELBACH, W. and ELFRIN, G. L., 1967. 'Structural and chemical asymmetry of the Ca-transporting membranes of the sarcotubular system as revealed by the E–M': *J. Ultrastruct. Res.* 17, 598.

66. HASSELBACH, W., 1964. 'Relaxation and the sarcotubular calcium pump': *Fed. Proc.* 23, 909.

67. HILL, A. V., 1966. 'A further challenge to biochemists': *Biochem. Z.* 345, 1.

68. HILL, A. V., 1953. 'The "instantaneous" elasticity of active muscle': *Proc. R. Soc. B*, 141, 161.

69. HILL, A. V., 1952. 'Thermoelasticity in muscle': *Proc. R. Soc. B*, 139, 464.

70. HILL, A. V., 1938. 'The heat of shortening and the dynamic constants of muscle': *Proc. R. Soc. B*, 126, 136.

71. HILL, A. V., 1949. 'The heat of activation and the heat of shortening in a muscle twitch': *Proc. R. Soc. B*, 136, 195–254.

72. HILL, A. V., 1958. 'The priority of the heat production in a muscle twitch': *Proc. R. Soc. B*, 148, 397.

73. HILL, A. V., 1950. 'Does heat production precede mechanical response in muscular contraction?': *Proc. R. Soc. B*, 137, 268.

74. HILL, A. V., 1964. 'The effect of load on the heat of shortening of muscle': *Proc. R. Soc. B*, 159, 297.

75. HILL, A. V. and HOWARTH, J. V., 1959. 'The reversal of chemical reactions in contracting muscle during an applied stretch': *Proc. R. Soc. B*, 151, 169.

76. HODGKIN, A. L., 1965. *Conduction of the Nervous Impulse:* Liverpool Univ. Press.

77. HODGKIN, A. L. and HUXLEY, A. F., 1952. 'Potassium leakage from an active nerve fibre': *J. Physiol.* 106, 341 and 449.

78. HOFMANN-BERLING, H., 1955. 'Geisselmodelle und ATP': *Biochim. biophys. Acta* 16, 146.

79. HOTTA, K., 1961. 'Model for the myosin ATP-ase active site': *Jap. J. Biochem.* 50, 218.

80. HUXLEY, H. E., 1953. 'Electron microscope studies on the organisation of filaments in striated muscle': *Biochim. biophys. Acta* 12, 387.

81. HUXLEY, H. E., 1963. 'Electron microscope studies of natural and synthetic protein filaments from striated muscle': *J. molec. biol.* 7, 281.

82. HUXLEY, H. E., BROWN, W. and HOLMES, K. C., 1965. 'Constancy of axial spacings in frog sartorius muscle during contraction': *Nature Lond.* 206, 1358.

83. HUXLEY, H. E. and BROWN, W., 1967. 'X-ray diffraction studies of muscle': *J. molec. biol.* 30, 383.
84. HUXLEY, H. E., 1964. 'Evidence for continuity between the central elements of the triads and extracellular space in frog sartorius muscle': *Nature Lond.* 202, 1067.
85. HUXLEY, A. F. and TAYLOR, R. E., 1958. 'Local activation of striated muscle fibres': *J. Physiol.* 144, 426.
85*a*. HUXLEY, A. F., 1957. 'A Theory of muscular contraction': *Prog. in Biophys. and Biophys. Chem.* 7, 255.
86. INFANTE, A. A., KLAUPIKS, D. and DAVIES, R. E., 1965. 'PC consumption during single working contractions of isolated muscle': *Biochim. biophys. Acta* 94, 504.
87. INFANTE, A. A., KLAUPIKS, D. and DAVIES, R. E., 1964. 'Length, tension and metabolism during short isometric contractions of frog sartorius muscles': *Biochim. biophys. Acta* 88, 215.
88. INFANTE, A. A., KLAUPIKS, D. and DAVIES, R. E., 1964. 'ATP: changes in muscles doing negative work': *Science* 144, 1577.
89. JEWELL, B. R. and WILKIE, D. R., 1960. 'Mechanical properties of relaxing muscle': *J. Physiol.* 152, 30.
90. JÖBSIS, F. F. and O'CONNOR, M. J., 1966. 'Calcium release and reabsorption in the sartorius muscle of the toad': *Biochem. biophys. Res. Commun.* 25, 246.
91. JÖBSIS, F. F., 1966. 'Force, shortening and work in relation to chemical changes during contraction': in *Symposium on Muscle* (Hungarian Acad. Sci., Budapest – in press).
92. JÖBSIS, F. F. and DUFFIELD, J. C., 1967. 'Force, shortening and work in muscular contraction: relative contributions to overall energy utilization': *Science* 156, 1388.
92*a*. JOHNSON, P., HARRIS, C. I. and PERRY, S. V., 1967. '3-methyl histidine in actin': *Biochem J.* 105, 361.
93. JOSEPHS, R. and HARRINGTON, W. F., 1966. 'Studies on the formation and physical chemical properties of synthetic myosin filaments': *Biochemistry* 5, 3474.
94. KAMMER, B. and BELL, A. L., 1966. 'Myosin filamentogenesis: effects of pH and ionic concentration': *J. molec. biol.* 20, 391.
95. KATZ, B., 1966. *Nerve, muscle and synapse:* McGraw-Hill and Co. Inc. – New York.
96. KITAGAWA, S., CHIANG, K. K. and TONOMURA, Y., 1964. 'Binding of p-nitrothiophenol to myosin A': *Biochim. biophys. Acta* 82, 83.
97. KOMINZ, D. R., HOUGH, A., SYMONDS, P. and LAKI, K., 1954. 'The amino-acid composition of actin, myosin, tropomyosin and the meromyosins': *Arch. Biochem. Biophys.* 50, 148.
98. KOSHLAND, D. E. (Jr.) and LEVY, H. M., 1964. 'Evidence for an intermediate in ATP hydrolysis by myosin': in *Biochemistry of Muscle*

Contraction, ed. J. Gergely, Little Brown and Co. – Boston, Mass., p. 87.

99. KUBO, S., TOKURA, S. and TONOMURA, Y., 1960. 'On the active site of myosin-A ATP- ase': *J. biol. Chem.*, 235, 2835.

100. KUSHMERICK, M. J., MINIHAN, K. and DAVIES, R. E., 1965. 'Changes in free P_i and ATP in frog sartorius muscles during maximum work and rigor': *Fed. Proc.* 24, 2.

101. LAUFFER, M. A., 1964. 'Protein–protein interaction: endothermic polymerisation and biological processes': *Symposium on Foods: Proteins and their interactions*, ed. H. W. Schultz. Avi Publishing Co. Inc. – Westport, Conn., p. 87.

102. LOWEY, S., GOLDSTEIN, L. and LUCK, S., 1966. 'Isolation and characterization of a helical sub-unit from heavy meromyosin': *Biochem. Z.* 345, 248.

103. LOWEY, S. and COHEN, C., 1962. 'Studies on the structure of myosin': *J. molec. biol.* 4, 293.

104. LUNDSGAARD, E., 1930. 'The energetics of anaerobic muscle contraction': *Biochem. Z.* 233, 322.

105. MAKINOSE, M., 1966. 'Die Nukleosidtriphosphat-Nucleosiddiphosphat-Transphosphorylase-Aktivität der Vesikel des Sarkoplasmatischen Reticulums': *Biochem. Z.* 345, 80.

106. MARÉCHAL, G., 1964. 'Le métabolisme de la phosphoryl-créatine et de l'adénosine triphosphate durant la contraction musculaire': Thése d'agrégation. Bruxelles. Editions: Arscia S.A.

107. MARECHAL, G. and MOMMAERTS, W. F. H. M., 1963. 'The metabolism of phosphocreatine during an isometric tetanus in the frog sartorius muscle': *Biochim. biophys. Acta* 70, 53.

108. MARECHAL, G. and BECKERS-BLEUKX, G., 1965. 'ATP and PC breakdown in resting and stimulated muscles after treatment with FDNB': *J. Physiol. (Paris)* 57, 652.

109. MARSH, B. B., 1952. 'The effects of ATP on the fibre volume of a muscle homogenate': *Biochim. biophys. Acta* 9, 247.

110. MARTONOSI, A., 1964. 'Role of phospholipids in ATP-ase activity and Ca transport of fragmented sarcoplasmic reticulum': *Fed. Proc.* 23, 913.

111. MEYERHOF, O. and SCHULTZ, W., 1927. 'Relation between lactic acid formation and O utilization in the contraction of muscle': *Pflügers Arch. ges. Physiol.* 217, 547.

112. MILLMANN, B. M., ELLIOTT, G. F. and LOWY, J., 1967. 'X-ray diffraction studies': *Nature Lond.* 213, 356.

113. MOMMAERTS, W. F. H. M., SERAYDARIAN, K. and MARECHAL, G., 1962. 'Work and chemical change in isotonic muscular contractions': *Biochim. biophys. Acta* 57, 1.

114. MUELLER, H. and PERRY, S. V., 1962. 'The degradation of heavy

meromyosin by trypsin': *Biochem. J.*, 85, 431. See also: JONES, J. M. and PERRY, S. V., 1966. 'Biological activity of Sub-fragment 1': *J. Biochem.* 100, 120.

115. MUELLER, H., 1966. 'EGTA-sensitizing activity and molecular properties of tropomyosin prepared in presence of an SH-protecting agent': *Biochem. Z.* 345, 300.

116. NANNINGA, L. B. and MOMMAERTS, W. F. H. M., 1960. 'Formation of an enzyme-substrate complex between myosin and ATP': *Proc. Nat. Acad. Sci. U.S.A.* 46, 1166.

117. NANNINGA, L. B., 1959. 'Investigation of the effect of Ca ions on the splitting of ATP by myosin': *Biochim. biophys. Acta* 36, 191.

118. NEEDHAM, D. M., 1960. 'Biochemistry of muscular action': *Structure and Function of Muscle*, ed. G. H. Bourne, vol. II, Chapter II.

118*a*. OFFER, G. W., 1966. 'N-acetyl peptides from myosin': *Biochim. et biophys. Acta* 90, 193.

119. PAGE, S., 1964. 'Sarcomere lengths in striated muscle': *Proc. R. Soc. B* 160, 460. 1967, Private communication.

120. PAUL, J., 1965. *Cell Biology*, Heinemann and Co. – London, pp. 35–47.

121. PEPE, F., 1967. 'The myosin filament: structural organization from antibody staining observed in electron microscopy': *J. molec. biol.* 27, 203.

122. PERRY, S. V., 1967. 'The structure and interactions, of myosin', *Progress in Biophys.* 17, 325. See also: SCHAUB, M. E., HARTSHORNE, D. J. and PERRY, S. V., 1967. *Nature Lond.* 215, 635.

123. PERRY, S. V. and GREY, T. C., 1956. 'A study of the effects of substrate concentration and certain relaxing factors on the Mg-activated myofibrillar ATP-ase': *Biochem. J.* 64, 184.

124. PODOLSKY, R. J., 1964. 'The maximum sarcomere length for contraction of isolated myofibrils': *J. Physiol.* 170, 110.

125. PODOLSKY, R. J. and MORALES, M. F., 1958. 'The enthalpy change of ATP hydrolysis': *J. biol. Chem.*, 218, 945.

126. PORTER, K. R. and PALADE, G. E., 1957. 'Studies on the endoplasmic reticulum III. Its form and distribution in striated muscle cells': *J. Biophys. Biochem. Cytol.* 3, 269.

127. PORTZEHL, H., SCHRAMM, G. and WEBER, H. H., 1950. 'Actomyosin and its components': *Z. Naturf.* 5B, 61.

128. PORTZEHL, H., 1957. 'Bewirkt das System Phosphokreatin-Phosphokinase die Erschlaffung des lebenden Muskels?': *Biochem. biophys. Acta* 24, 474.

129. PORTZEHL, H., CALDWELL, P. C. and RÜEGG, J. C., 1964. 'Calcium and the contraction of crab muscle fibres': *Biochem. biophys. Acta* 79, 581.

130. PRINGLE, J. W. S., 1967. 'The contractile mechanism of insect fibrillar muscle': *Progress in Biophys.* 17, 1. See also *Am. zool.* 7, 465 (1967).

131. REVEL, J. P., 1964. 'The sarcoplasmic reticulum of fast-acting muscle':

Biochemistry of Muscle Contraction, ed. J. Gergely, Little Brown and Co., Boston, Mass., p. 232.

132. REVEL, J. P. and FAWCETT, D. W., 1961. 'The sarcoplasmic reticulum of a fast-acting fish muscle': *J. Biophys. Biochem. Cytol.* Vol. 10, No. 4, pt. 2 (suppt.) 'The sarcoplasmic reticulum', p. 89.
133. RICE, R. V., BRADY, A. C., DEPUE, R. H. and KELLY, R. E., 1966. 'Morphology of individual macro-molecules and their ordered aggregates by electron microscopy': *Biochem. Z.* 345, 370.
134. SANDBERG, J. A. and CARLSON, F. D., 1966. 'The length dependence of PC hydrolysis during an isometric tetanus': *Biochem. Z.* 345, 212.
135. SANDOW, A., 1947. 'Latency relaxation and a theory of muscular mechano-chemical coupling': *Ann. N.Y. Acad. Sci.* 47, 895.
136. SANDOW, A., TAYLOR, S. R. and PREISER, H., 1965. 'Role of the action potential in excitation-contraction coupling': *Fed. Proc.* 24, 1116.
137. SANDOW, A., 1965. 'Excitation-contraction coupling in skeletal muscle': *Pharmacol. Rev.* 17, 265.
138. SARTORELLI, L., FROMM, H. J., BENSON, R. W. and BOYER, P. D., 1966. 'Direct and ^{18}O-exchange measurements relevant to possible activated or phosphorylated states of myosin': (in press).
139. SEIDEL, J. C., SRETER, F. A., THOMSON, M. M. and GERGELY, J., 1964. 'Comparative studies of myofibrils, myosin and actomyosin from red and white rabbit skeletal muscle': *Biochem. biophys. Res. Commun.* 17, 662. See also: GERGELY, J., PRAGAY, D., SCHOLZ, A. F., SEIDEL, J. C., SRETER, F. A. and THOMPSON, M. M. in *Molecular Biology of Muscular Contraction*, Igaku Shoin Ltd, Tokyo, 1965, p. 145.
140. SHOENBERG, C. F., RÜEGG, J. C., NEEDHAM, D. M., SCHIRMER, R. H. and NEMETCHEK-GANSLER, H., 1966. 'A biochemical and electron microscope study of the contractile proteins in vertebrate smooth muscle': *Biochem. Z.* 345, 255.
141. SISSON, S., 1927. *The anatomy of the domestic animals*, 2nd ed., W. B. Saunders and Co. – Philadelphia.

141*a*. SLAYTER, H. S., and LOWEY, S., 1967. 'The sub-structure of the myosin molecule as visualised by the electron microscope': *Proc. Nat. Acad. Sci. U.S.A.* 58, 1611.

142. STRACHER, A., 1965. 'Evidence for histidine at the active site of myosin A': *J. biol. Chem.* 240 P.C. 958.

142*a*. STRACHER, A., 1964. 'Disulfide-sulfhydryl interchange studies on myosin A': *J. Biol. Chem.* 239, 1118.

143. STRAUB, F. B., 1942. 'G- and F-actin and effect of ATP': *Studies from Inst. Med. Chem. Univ. Szeged.* 2.3, quoted by A. Szent-Györgyi (q.v.).
144. SZENT-GYÖRGYI, A., 1945. 'Studies on muscle': *Acta Physiol. Scand.* 9, Suppt. XXV.
145. SZENT-GYÖRGYI, A., 1953. *Chemical physiology of contraction in body and heart muscle.* Acad. Press Inc. N.Y.

146. SZENT-GYÖRGYI, A. G., 1960. 'Proteins of the myofibril': *Structure and Function of Muscle,* vol. II, chapter I, ed. G. H. Bourne, Academic Press – New York.

146*a*. SZENT-GYÖRGYI, A. G. and PRIOR, G., 1966. 'Exchange of ADP bound to actin in superprecipitated actomyosin and contracted myofibrils': *J. molec. biol.* 15, 515.

147. TRAYER, J. C., 1966. 'Studies on myosin ATP-ase in developing muscle': Ph.D. Thesis – University of Birmingham.

148. TRAYER, I. P. and PERRY, S. V., 1966. 'The myosin of developing skeletal muscle': *Biochem. Z.* 345, 87.

149. TONOMURA, Y., 1965. 'A molecular model for the interaction of myosin with ATP': in *Molecular Biology of Muscular Contraction,* ed. S. Ebashi, F. Oosawa, T. Sekine and Y. Tonomura, Igaku-Shoin Ltd, Tokyo, p. 11.

150. ULBRECHT, G., ULBRECHT, M. and WUSTROW, H. J., 1957. 'Beruht der Phosphat-Austasch zwischen ATP und $AD^{32}P$ durch hochgereinigte Aktomyosin-Präparate': *Biochim. biophys. Acta* 25, 110.

151. VERATTI, E., 1902. 'Investigations on the fine structure of striated muscle fibres': reprinted in *J. Biophys. Biochem. Cytol.* Vol. 10, No. 4, pt. 2 (suppt.), 1966, p. 1.

152. WATANBE, S. and SLEATOR, W., 1957. 'EDTA relaxation of glycerol-treated fibres and the effects of Mg, Ca and Mn ions': *Arch. Biochem. Biophys.* 68, 81.

153. WEBER, A. and WINICUR, S., 1961. 'The role of calcium in the superprecipitation of actomyosin': *J. biol. Chem.* 236, 3198.

154. WEBER, A. and HERZ, R., 1963. 'The binding of calcium to actomysin systems in relation to their biological activity': *ibid.* 238, 599.

155. WEBER, A. and WEBER, H. H., 1951. 'Zür Thermodynamik der Kontraktion des Fasermodelle': *Biochim. biophys. Acta* 7, 339.

156. WEBER, A., HERZ, R. and REISS, I., 1963. 'On the mechanism of the relaxing effect of fragmented sarcoplasmic reticulum': *J. gen. Physiol.* 46, 679.

157. WEBER, A., HERZ, R. and REISS, I., 1966. 'Study of the kinetics of Ca transport by isolated fragmented sarcoplasmic reticulum': *Biochem. Z.* 345, 329.

158. WEBER, A., HERZ, R. and REISS, I., 1964. 'The regulation of myofibrillar activity by Ca': *Proc. R. Soc. B,* 160, 489.

159. WEBER, H. H. and PORTZEHL, H., 1954. 'The transference of the muscle energy in the contraction cycle': *Prog. Biophys. and Biophys. Chem.* 4, 91.

160. WEBER, H. H., 1958. *Motility of Muscle and Cells.* Harvard Univ. Press, Cambridge, Mass. See also, YOUNG, L. G. and NELSON, L., 1968. *J. Exptl. Cell Res.* 51, 34.

161. WILKIE, D. R., 1956. 'Measurement of the series elastic component at various times during a single muscle twitch': *J. Physiol.* 134, 527.

162. WILKIE, D. R., 1962. 'Tissues subserving movement and conduction': in *Principles of Human Physiology* by Starling and Lovatt Evans, 13th edition, J. and A. Churchill, London, pp. 794–843.
163. WILKIE, D. R., 1960. 'Thermodynamics and the interpretation of biological heat measurements': *Prog. Biophys. and Biophys. Chem.* 10, 259–98.
164. WINTON, F. R. and BAYLISS, L. E., 1962. 'Nerve, neuromuscular and synaptic transmission, and the central nervous system': *Human Physiology*, 5th edition, Churchill Ltd, London, 346–90.
165. WOLEDGE, R. C., 1963. 'Heat production and energy liberation in the early part of a muscular contraction': *J. Physiol.* 166, 211. See also ibid: 1968, 197, 685.
166. YOUNG, M. D., HIMMELFARB, S. and HARRINGTON, W. F., 1965. 'On the structural assembly of the polypeptide chains of heavy meromyosin': *J. biol. Chem.* 240, 2428.
167. REES, M. K. and YOUNG, M., 1967. 'The structure and properties of G-actin': *J. biol. Chem.* 242, 4449.
168. KIELY, B. and MARTINOSI, A., 1968. 'Kinetics and substrate binding of myosin ATP-ase': *J. biol. Chem.* 243, 2273.
169. ALBERTY, R. A., 1968. 'Equilibrium hydrolysis of ATP and the binding constants of MgATP, CaATP, MgADP and CaADP': *J. biol. Chem.* 243, 1337.
170. MCNEILL, P. A. and HOYLE, G., 1967. 'Evidence for superthin filaments in muscle': *American Zoologist.* 7, 483 *et seq.*
171. GREEN, D. E., ASAI, J., HARRIS, R. A. and PENNISTON, J. T., 1968. 'Confirmational basis of energy transformations in membrane systems: III': *Arch. biochem. and biophys.* 125, 684.
172. WEEDS, A. G., 1969. The light chains of myosin: *Nature,* 223, 1362.
173. FINLAYSON, B. and TAYLOR, E. H. (1969). Hydrolysis of Nucleoside triphosphates by myosin during transient state. *Biochemistry*, 8, 802 (see also p. 811).

For general reference to the physiology of living muscle, consult:

A. V. Hill, 1965. *Trails and Trials in Physiology*, Edward Arnold (Publishers) Ltd., London (1965).

Index